# Under the Weather

# Under the Weather

## How the Weather and Climate Affect our Health

PAT THOMAS

ff

First published in 2004 by Fusion Press,
a division of Satin Publications Ltd.
101 Southwark Street
London SE1 0JF
UK
info@visionpaperbacks.co.uk
www.visionpaperbacks.co.uk
Publisher: Sheena Dewan



A catalogue record for this book is available
from the British Library.

ISBN: 1-904132-30-8

2 4 6 8 10 9 7 5 3 1

Cover and text design by ok?design.
Printed and bound in the UK by
Mackays of Chatham Ltd, Chatham, Kent.

*to weather watchers everywhere*

# Contents

# Acknowledgements

Although this book started out as a small project, it soon became apparent that there was a wealth of information on the weather/health link buried in various libraries, second-hand bookshops and medical and scientific journals around the world. Many things in my life were put on hold while I ferreted out these titbits. I am indebted to all the usual crowd who have been patient with me over the writing of this book, including (first and foremost) my son Alex. Thanks to Lynne McTaggert and Bryan Hubbard for their tolerance and encouragement, and all my friends whose kind invitations to 'chill out' with them I have had to decline while I remained tethered to my computer. My agent Laura Longrigg's faith is a continuing source of strength. Grateful appreciation to all at Fusion Press, including Charlotte Cole, for seeing the potential in this fascinating subject.

# Preface

Let's talk about the weather.

When I was in my late teens I moved from Los Angeles where we had climate to London where we have weather. I made my entrance on a cold November night in an unsuitable pair of slingback sandals, lugging a suitcase full of unsuitable clothes, and promptly fell down a flight of icy stairs onto what was then a bleak outdoor rail platform attached to Gatwick airport.

A very sensibly dressed couple in practical shoes picked me up, dusted me off and asked if perhaps I had any more appropriate footwear that I could change into before continuing my journey.

I fell over a lot that winter. It was the first time I had ever been exposed to extreme cold – never mind ice and snow – on a day-in, day-out basis. Plugging endless coins into the two-bar electric fire in my rented flat, I felt quite literally sick at heart at the monumental miscalculation I had made in coming to live in such an inhospitable part of the world.

Learning to adapt to the weather can be difficult; yet if human evolution has taught us anything, it is that we are masters at adaptation.

Humans have thrown considerable energy and ingenuity at adapting to and protecting themselves against the weather. We build shelters. We engineer and wear seasonal clothing. We damp- and windproof our homes. In winter, we insulate and heat; in summer we switch on the fans and turn up the cold air in our indoor environments. We have, in fact, become so good at adapting to the weather that we have forgotten just what a powerful force it is, and that it can influence our health and well being more deeply than cold toes and chapped lips.

While some might argue that all the things we do to protect ourselves from the weather make us immune to its effects, there is another more persuasive argument to suggest that all the trappings of modern life actually make us more susceptible to abrupt changes in weather. Walking from a highly-heated home into the cold of a winter's day can mean experiencing an almost immediate drop in temperature of 10° C/50° F – the same kind of sudden temperature change that has been shown to increase the risk of cardiovascular failure in vulnerable individuals.

Our gradual detachment from nature and our modern lifestyle – complete with air conditioners, humidifiers and central heating – has made us more sensitive to environmental influences. Our bodies have lost much of the adaptability that would otherwise help them cope with the shock of rapidly changing weather elements. If you are fit and healthy, the shock may remain unnoticed, or pass as a minor nuisance. But if your body is weakened by stress or age, the result can be a range of mood disorders, illness and even death.

What is more, scientists predict that our game of one-upmanship with nature, played by releasing greenhouse gases and pollution into the environment, may ultimately make our sensitivity to the weather even more acute. Global warming is expected to increase the likelihood of severe weather all over the world, making it less predictable, more extreme and, for a variety of reasons, more damaging to human health.

To some, such statements may seem like quackery or an archaic form of armchair meteorology. If weather was really important to health, surely our doctors and scientists would have told us so? Unfortunately, traditional scientific study, by its very nature, factors out as unimportant anything that man cannot 'control'. So it is not surprising that the weather, and its diverse, and occasionally subtle, influences on human health, have been sidelined in favour of more predictable disease-causing agents such as germs and genetic theory.

Many of us realise on an unconscious level that weather affects the way we feel. Why else would we refer to ourselves as being 'under the weather' when we are ill? Why else would so many of us attribute sudden changes in our health or well being to changes in the weather? What most of us don't realise is the extent to which weather controls our health.

A growing body of scientific evidence is confirming the weather/health link and suggests that around 1 in 3 of us are 'weather sensitive'. Increasingly, researchers throughout the world are now using weather and climate predictions as a compass to track health trends and develop preventive measures. This relatively new scientific discipline is known as 'Biometeorology'. On a very large scale, biometeorologists look at the relationship between atmospheric patterns around the world, changes in climate and effects on global health patterns. On a smaller scale, they study the effects of weather on individual health.

There are stalwart scientists who are piecing together this complex information, and their findings are fascinating, not only on an intellectual level, but also on an intuitive one. That our bodies could interact with the natural environment at such a deep level shows just how much a part of nature we are. It confirms that the human organism is extraordinarily sensitive to its surroundings – more so than most of us realise. The weather, with its mixture of electrical forces, heat, water and wind, is a natural influence that defines and controls us. Try as we might to separate who we are, how we behave and the state of our health from environmental influences, we cannot.

Most of us talk about the weather when we want to pass the time in a bland sort of way. If this book does anything at all, it will hopefully lift the curtain on weather as a dynamic force that shapes and influences our lives in unexpected and profound ways.

Next time you say to someone 'It looks like rain', it could just be the entrée to a whole different sort of conversation and an entirely different way of thinking.

*Pat Thomas*

# Chapter One

## The Human Barometer

> **Whoever wishes to pursue properly the science of medicine must proceed thus. First, he ought to consider what effects each season of the year can produce: for the seasons are not alike, but differ widely both in themselves and at their changes...**[1]

This was the advice offered to future physicians by Hippocrates, the founder of modern medicine. In his book, *Air, Waters and Places*, written some 2,300 years before any modern scientific studies were done on the subject, he was perhaps the first scientist to document his theories and observations about how the weather affects our well being.

Since the beginning of our reign on earth, humans have been more often exposed to the elements than protected from them. Much like other animals, our bodies have evolved to work with prevailing weather conditions rather than against them. As one modern author put it 'Because 98 per cent of our skin is in intimate contact with the ocean of air that engulfs us – at once our ancestral blanket, home, and master – the atmosphere conditions humanity.'[2]

The influence of the weather is considered vital to the understanding of human health in many ancient healthcare systems such as traditional Chinese medicine and Ayurveda. Homeopathic practitioners take into account, among other things, if certain weather conditions make a disorder better or worse, and select their remedies accordingly. Most (but not all) conventional medics reject such views, believing that the likelihood of health or illness rests entirely on the twin horns of germs and genetics.

Yet in some parts of the world, weather is widely acknowledged as a contributing factor in many and varied health conditions. For more than a decade, German physicians have been able to make use of daily bulletins from the national weather service, Deutscher Wetterdienst, to help them advise patients on the management of common health problems. Following this lead, the national weather bureaus in the US and the UK also now issue bulletins that go well beyond advising hay-fever sufferers of high pollen counts.

In the US, health-conscious web surfers can log on to the health pages at the Internet weather channel Intellicast.com and check their health forecast – for instance, the likelihood of aches and pains, respiratory problems, mood swings and changes in attentiveness and reaction times – in the same way they might check the news or the stock market. Pregnant women can even gauge the likelihood of their going into labour. These pages get more than 100,000 hits every month.

In the UK, the official weather agency, the Met Office, is currently developing an early-warning system, based on changes in the weather, that will help hospitals better predict how many patients they will have to treat. The system has already been tested at five locations, and is currently being tested at another 30 hospitals around the country.[3]

Although the system is still evolving, the current prototype collects data on temperature, humidity, barometric pressure and predicted precipitation as well as monitoring seasonal diseases such as influenza. This data is used to give hospitals and family doctors advance notice of increases in illnesses that are linked to changes in the weather, such as heart attacks, strokes, respiratory illness, infectious disease and broken bones.

Early results showed that, in one of the test areas, one hospital was able to perform 150 extra operations due to the forecast predicting a lower emergency workload than usual, allowing extra beds to be used for elective surgery. Without the forecast, this would have been too risky because of a possible influx of emergency patients.

Weather is a dynamic force, always present and constantly changing. That the human body is sensitive to these changes is without question. What scientists today are trying to discover is the degree of and mechanism behind that sensitivity.

## What makes the weather?

Everyone knows what weather is. It's the rain that ruins the picnic, the sun that makes or breaks a holiday, the snow that holds up the traffic and the wind that messes up your hair and blows grit into your eyes. For most of us weather is something that can impact on the practical aspects of everyday life, but beyond that most of us have never wondered what shapes the weather's many faces.

Weather and climate are, of course, not the same things. It has been observed that while 'climate is what you expect, weather is what you get'. Weather can be defined as the day-to-day and in some cases hour-to-hour changes in atmospheric conditions. Climate is an average of those conditions over time. As one scientist puts it, '...climate suppresses or expands the range of many diseases, and weather contributes to [their] timing and intensity...thus, in the context of disease, climate provides motive, and weather provides opportunity'.[4]

Several factors contribute to what we experience as weather. These include atmospheric conditions such as temperature, wind speed, humidity and precipitation, but also geographical factors such as altitude, latitude, time of day and time of year.

But most scientists agree that the catalyst for all our weather experiences begins in the atmosphere – that cushion of air that extends from sea level to more than 62 miles/100 km above the earth's surface. Human life evolved at the bottom of this ocean of air, which is a rich mixture of gases including nitrogen (78 per cent), oxygen (20 per cent), and argon (less than 1 per cent) as well as minute quantities of neon, helium, krypton, hydrogen, xenon, ozone and radon.

Although we can't see it, the atmosphere performs a kind of ballet composed of a continuous cycle of rising warm air and falling colder air. This rising and falling is played out over a vast scale spanning oceans and continents, but also on a smaller scale, in cities and towns, all over the world. These areas of rising warm air and falling colder air are called low- and high-pressure areas.[5]

To understand the difference between high- and low-pressure areas, imagine the air being sucked up by a giant straw. Wherever warmer air is being drawn into the upper atmosphere, a type of vacuum is created near the surface of the earth. This is a low-pressure area. Likewise when cooler, denser air begins piling up or falling, an area of high pressure is created.

While the air in the upper part of the atmosphere moves

rapidly – 6 to 8 miles or 9.6 to 11.2 km above the earth, the 'jet stream', for instance, can reach speeds of up to 200 mph – closer to the earth's surface, air movements are slowed down because of friction with the earth. Wind also tends to move from areas of high pressure to areas of low pressure. But because the earth is continually spinning, the great air masses that rise from the equator and fall at the poles are deflected sideways in what is known as the Coriolis effect. All

## Warm and cold fronts

Every TV weather report includes some information on what are known as warm and cold 'fronts'. When they meet in the atmosphere, warm air and cold air don't immediately 'mix'. This is because air at different temperatures has a different mass and weight.

Fronts are masses of warm and cold air that crash into each other, often at great speed, causing rapid changes in air pressure and temperature. Think of the effect of a mini crashing into a four-wheel drive vehicle and you will get some idea of the impact when warm and cold fronts meet.

A warm front develops on the leading edge of a mass of warm air. Because cold air is heavier and more dense than warm air, the warm air mass is forced up over the top of the cooler one. This pushes it higher into the atmosphere, where it cools, water vapour condenses and clouds and precipitation are typically formed. As the air begins to rise, the atmospheric pressure, and thus the barometer, will gradually begin to fall.

Warm fronts are typically followed by cold fronts. When a dense, cold air mass slams into the back of a warm air mass, it violently and abruptly forces the warm air upward. When this happens, the already gradually falling barometer (the result of the passing warm front), falls sharply, dramatically lowering the atmospheric pressure. The rapid cooling of the rising air typically generates violent storms.

The arrival of warm and cold fronts signals the kind of changeable weather patterns that are associated with sometimes dramatic changes in our well being.

of this air movement creates what we experience as the wind.

Because of the Coriolis effect, the weather systems in the Northern Hemisphere typically move from west to east (in the Southern Hemisphere they move from east to west). Not all of the great masses of warm air rising from the equator are affected in this way, however. Some of the rising air slips away in northerly and southerly directions, thus reaching the poles. As the cold air gathers and sinks at the poles, it begins to cycle back towards the equator, and the dance begins again.

## Early investigations

The study of how the weather and climate affect humans has variously been called 'climatology', 'medical climatology', 'clinical climatology', 'bioclimatology', 'clinical biometeorology' and finally today just 'biometeorology'. It is fair to say that in some medical circles, biometeorology is denigrated as a kind of high-tech version of the *Farmers' Almanac* for the 21st Century; looking for a scientific basis for what some believe are merely old wives' tales, dubious horoscopes and crackpot theories.

But the scorn poured on biometeorololgists is not necessarily because their theories or their science is faulty. Instead, it is because they attempt what most scientists try hard to avoid – measuring the 'unmeasurable', embracing the 'changeable' and 'uncontrollable' and understanding the often-subtle physical effects of weather on the body. To do this, biometeorologists take into account the interrelationship of many complex factors, such as the combined effects of temperature, humidity, number of sunshine hours, wind speed, precipitation and geomagnetic activity generated by the earth, the sun and the moon.

Sceptics might be surprised to learn that, far from being a fringe science, biometeorology is a profession with university courses and several peer-reviewed journals to its credit. As a result, there has been a surprising amount of modern research into the myriad effects that changing weather has on the human body.[6]

In the early 1900s, American geographer Ellsworth Huntington published his book *Civilization and Climate*.[7] Huntington believed that the way that human progress ebbed and surged forward in waves was the result of an interaction between climate, quality of people and culture. His beliefs culminated in his own magnum opus *Mainsprings of Civilization*,[8] which contended (based on his own extensive studies of various populations across the US) that our intellectual and physical vigour depended to a large extent on being exposed to a climatic optima where temperature and humidity were concerned. Taking a worldwide view, he believed that place was more influential than race when it came to differences between cultures.

Unlike many biometeorologists that followed, Huntington believed that variability in climate was inherently healthy. Accordingly, he believed that for physical health, an average temperature of 18° C/64° F and a mean humidity of 80 per cent were best. For physical work, the best temperature was around 15.5° C/60° F and for mental work, 4.5° C/40° F.[9] 'The only way to get all these conditions,' he wrote, 'is to live in a climate which has several frosty but not cold months in the winter, several warm but not hot months in the summer, and a constant succession of storms at all seasons...Such conditions apparently prevailed in the region where our ancestors acquired their present adaptation to climate.'

Another early scientist who studied these effects was a Dr William F. Petersen who, in 1935, published his views in a book called, *The Patient and the Weather*.[10]

What we call cold fronts today, Petersen referred to as polar fronts. His own extensive research suggested that when a polar front approaches and the atmospheric pressure begins to fall, the body responds by contracting blood vessels and reducing the amount of oxygen to the heart, brain, kidneys

## Homeotherms

Human life evolved largely within areas where the climate was mild, or temperate. We are happiest close to the ground in stable areas of relative high pressure. Most of us feel quite comfortable with an air temperature of 25° C/77° F. A fully-clothed person indoors feels comfortable at between 20–25° C/68–77° F. Within this 'thermoneutral' range, the amount of heat lost from the body (through radiation or when we sweat – see Chapter 6) is equal to the amount gained (for example, from the sun) and the body does not have to work too hard to maintain its thermal balance.

Our bodies face considerable challenges from humidity, wind, ultraviolet radiation and variations in air pressure and temperature because humans are homeotherms, warm-blooded mammals that regulate their internal body temperature within a very narrow range. The average body temperature is around 37° C/98.6° F – it is potentially life threatening for the body's core temperature to vary from this by more than 2° C.

The body core is made up of the vital organs: brain, heart, lungs, kidneys and digestive tract. Enzymes within these organs assist in important biochemical reactions. These enzymes operate best in a temperature range of 35–40° C/95–104° F. Outside of this range, they may undergo loss of function or death and bodily functions may become impaired due to the loss of this enzyme activity.

and other major organs. This causes blood pressure to rise, resulting in an overall 'stimulation' of the body.

According to Petersen, to counter the reduced oxygen availability in the body, the call goes out to release substances into the blood to dilate the blood vessels, restore normal blood flow and lower blood pressure. Petersen believed that such changes accounted for the way that some people could appear upbeat one day, and relatively sluggish the next. Indeed, he was the man who coined the oft-used phrase 'cosmic resonator' to describe man's relationship with his environment.

In the 1960s, S.W. Tromp, in his book *Biometeorology: The Impact of the Weather and Climate on Humans and Their Environment*, echoed many of Petersen's claims about the body's reaction to atmospheric pressure changes. Much of his research was done on native populations living at high or low elevations (and consequently at low or high atmospheric pressure conditions), and also on subjects in pressure chambers.[11]

According to Tromp, when our bodies are exposed to a low ambient pressure, several largely involuntary physiological changes, including an increase in our breathing rate and blood pressure, take place.

While Tromp agreed that some of these changes occurred with a drop in atmospheric pressure, he did not believe that changes in air pressure were the only weather triggers for changes in our well being. His own, indeed his first, fascination was with the way geomagnetic forces shape our health and behaviour or, as he put it, the 'wonderful web of electromagnetic forces which seem to regulate all living processes on earth'.[12] While it has not been updated in more than 20 years, his careful, often eloquent observations and insights have ensured that Tromp's *Biometeorology*, remains a standard text in the field.

A number of cultural changes in the 1960s raised the bar on our understanding of ourselves as both impacting on and impacted by the natural environment. Around the same time that Tromp's work was making its influence felt, a contemporary of his, meteorologist Helmut E. Landsberg PhD, research professor at the University of Maryland and Chairman of the Graduate Committee on Meteorology, developed a simple yet sophisticated system for classifying weather patterns as they approach and leave local areas. Landsberg's weather cycle consisted of six distinct phases, each characterised by consistent, factual, weather occurrences.[13]

**Phase I** high pressure, cool temperatures, light to moderate winds, low humidity, few clouds.

**Phase II** high pressure, light winds, sunny clear skies.

**Phase III** slightly falling pressure, high clouds moving in.

**Phase IV** falling pressure, rising temperature and humidity, thickening clouds, rain or snow.

**Phase V** rapidly rising pressure and falling temperature, falling humidity, gusty winds.

**Phase VI** slowly rising pressure, low temperature and humidity, slackening winds.

Using this system allowed Landsberg to collect and compare data on a variety of different human behaviours as they related to prevailing weather conditions.

During one of his studies, 20,000 visitors to a traffic exhibition in Germany were given reaction-time tests over a ten-week period. Using his own system of weather phases for comparison, he found that reaction times were quicker during phases I, II and VI and slower during phases III, IV and V.

Landsberg also observed that more industrial accidents occurred during phases III, IV and V and that people tended to be more irritable during periods of sustained temperature increases. His studies led him to believe that while changes in barometric pressure were important, these were, in the end, '...merely an index for the whole system of weather patterns'.

Landsberg's studies also encompassed the electromagnetic fields (EMFs) produced by thunderstorms. As barometric pressure falls rapidly during the passing of a cold front, the likelihood of thunderstorms increases. When they do erupt, there is an overall increase in electromagnetic activity in the surrounding atmosphere. Landsberg, like Tromp, believed that there was a definite interaction between fluctuating natural electric fields, human brain-wave patterns and human behaviour.

## New directions

It would be easy to dismiss such observations and studies, some of which took place more than 50 years ago, as old fashioned and irrelevant to modern living. But research into weather effects on human health continues today. Far from disproving the findings of early pioneers, it appears to validate them.

Dr. Michael Persinger of Laurentian University, Sudbury, Ontario is one of today's most prominent biometeorologists, and a supporter of Landsberg's theories. With a background in neuroscience and clinical psychology, Persinger has done thousands of tests with electromagnetic waves, similar to those generated by storms, and the reaction of the human brain.

Though his work is not without its critics, Persinger believes electromagnetic waves have a profound influence on the way human cells function. Naysayers believe that the types of waves associated with storms are too low in intensity to affect humans (a claim often made in support of the safety of mobile phones and other electrical gadgets). But Persinger maintains it is not the waves' intensity, but their frequency and shape that are important.

Sooner or later, the nervous system – the bundles of nerves and fibres that control our bodies' internal workings – will respond to the faint but persistent barrage. Persinger likens it to trying to talk to someone in a noisy room. You can shout, but the likelihood is that the person you are trying to talk to will become so frustrated that they will tune you out altogether or just leave. But if you whisper, it forces the person to pay attention.

Whispers, says Persinger, cut through the background noise because they carry two important messages: the underlying or metamessage, which is 'this is important' and also the actual message 'do this'. His belief is that low-frequency EMFs grab the brain's attention in much the same way.

His view is the result of his own extensive research. By applying faint electromagnetic pulses to the temporal lobes of his volunteer subjects' brains, Persinger has been able to produce a variety of different moods from mild euphoria to depression.

Persinger's view, of course, challenges the status quo, and because of this he has often had trouble finding funding. Like many dedicated scientists, he has often been forced to fund his experiments out of his own pocket.[14]

Today's biometeorologists are divided as to how far they should interpret their findings. In order to facilitate acceptance of this science, many prefer to stick to the 'facts' – for instance, the way that extremes of heat and cold can kill – and leave interpretation of more esoteric data to someone else. Yet it is hard to look at the accumulated data on the interaction of humans with their weather environment and not feel that some of the discoveries of biometeorologists have important implications for human health.

For instance, in the late 1990s, one of Persinger's students, Rod O'Connor, found that the incidence of sudden infant death syndrome appears to be higher when geomagnetic storm activity is very low.[15] There is evidence that infants who die of SIDS have low levels of melatonin, a hormone that helps to control our internal clocks but which also mediates the production of nitric oxide, a chemical transmitter that regulates breathing.

O'Connor's theory is that very low-level magnetic fields generated during weather events depress nocturnal levels of melatonin and thus nitric oxide, resulting in a disruption of the infant's breathing (it is interesting to note that in the UK another scientist, Roger Coghill, has also found evidence that low level EMFs – this time man-made – correlate with a higher incidence of SIDS).[16]

Officialdom has largely ignored such findings and probably would not know how best to advise parents even if they did take them on board. The advice to place babies on their backs (and most recently not to let them overheat), while it does not explain the cause of this disorder, appears to work and is seen as a practical solution that the public can easily take on board without having to think too hard. It does not, however, advance our understanding of what is more than likely a complex syndrome that has several contributing factors.

But, speaking in a recent magazine article,[17] Persinger defended biometeorologists' right to extrapolate. Biometeorology matters, he argued, because the way in which we order out lives in the 21st century leaves us especially vulnerable to weather insults. 'Our indoor environments have become extraordinarily physically constant. The lighting. The temperature of our houses. One thing we do know about biological systems is that they tend to respond to contrast. And one of the things that fluctuates most these days is the weather.'

## Weather sensitivity

The dismissal of the weather's influence by the mainstream of Western thought is a relatively recent development.

Although Mark Twain once opined, 'Everybody complains about the weather, nobody does anything about it', this is not strictly true. Since he first walked the earth, man has striven to shield himself from extremes in weather and invent better methods of indoor 'climate control'. Advances in science and technology have helped create the illusion that we have, in fact, mastered the weather. It is only when faced with weather's extremes – hurricanes, tornadoes, killer heatwaves – that we are prodded into acknowledging that nature still has the upper hand.

Our dismissal of the weather also has its roots in the 'Enlightenment' of the 18th century, when scientific thought replaced Romantic idealism. Then, as now, science demanded exact rigorous measurements taken in totally controlled environments. Then, as now, anything that could not be controlled was taken out of the equation. Hence physicians began to conduct their experiments in chambers free of atmospheric turbulence. Science, in effect, moved indoors.[18]

Such thinking was in line with other emerging beliefs, such as man's superiority over the rest of nature. The Industrial Revolution moved people from farms to factories and from rural communities into crowded cities, resulting in the almost total repression of the influence of nature and of our natural inclinations.

We can lessen the impact of weather on our day-to-day lives. But weather could not, and still cannot, be controlled. Therefore modern thinking has gone the way of ignoring its influence altogether. Nevertheless, nearly every culture has preserved some perception of weather's importance to everyday life.

The worship of atmospheric powers is, for instance, difficult to separate from the worship of heaven. It is hardly surprising, then, that in many belief systems and mythologies the high god in heaven is also a weather god. The Norse god Thor ruled thunder. In Vedic myths, it is Indra who is supreme god and lord of thunder and lightning. For the Celts, Lugh was the sun god and protector of the crops, and Japanese Buddhists worshipped Nikko-Bosastu, god of sunshine and good health.

Weather proverbs, handed down from generation to generation, may seem like nothing more than nonsense rhymes. Yet some, especially those that 'forecast' short-term weather changes, are based in careful observation of nature.[19] Take, for instance, 'Mackerel skies and mares' tails, make tall ships carry low sails.' 'Mackerel' and 'mare's tail' are early descriptions of low alto-cumulus and cirrus clouds – typical of an approaching warm front and possible rain. Similarly, 'Sound travelling far and wide, a stormy day will betide' reflects what we now know to be the way that lighter water molecules in the humid, pre-storm environment increase the velocity at which sound travels through the air.

The modern obsession with collecting statistics suggests that ancient impressions of how the weather can affect our bodies and especially our minds are also often correct. In Spain it is said that the highest suicide rates are found in a town at the southern tip, constantly bombarded by wind from the Atlantic. Arctic dwellers have higher rates of depression and suicide, while in the hot and humid southern states of America, rates of crime and aggression rise with the mercury – though these are often attributed to what has come to be known as the 'Southern culture of violence'.

Recent evidence from Germany suggests that more than 50 per cent of individuals there are weather sensitive.[20] Some argue that the mere fact that weather sensitivity is taken seriously in that country means that the concept becomes a self-fulfilling prophecy. This however seems a little harsh, especially since a more recent survey by the Weather Channel in the US, where weather health bulletins are less common, reputedly found that around 70 per cent of Americans believe that the weather has an influence on their well being.[21]

We are all susceptible to suggestion at some level. But another way of looking at such statistics is that in a culture where weather sensitivity is taken seriously more people will feel at ease admitting that the weather affects them.

### Smelling rain

People with arthritis often comment, 'I can feel a storm coming.' This kind of armchair weather prediction has fallen into legend. But when, in 1961, Dr J.L. Hollander of the Graduate Hospital of the University of Pennsylvania placed some volunteers with arthritis in isolated hospital rooms where the atmospheric conditions could be altered, he found that whenever the pressure fell and humidity increased – conditions similar to those of an approaching storm – the patients did actually complain of pain in their joints. When the patients were examined, it was evident that their joints were truly inflamed. Although the mechanism is not well understood, it is now accepted that changes in air pressure and humidity do affect joints.[22]

Other people believe that they can detect the approach of a storm by the smell of the air. The 'smell' of rain during and after a rainstorm is probably caused by rainwater pushing gases, created by certain bacteria, out of the soil. These gases have their own peculiar odour. But the reasons why some individuals can smell rain before it falls are less clear.

In part, it may be that an increase in moisture and warmth and a decrease in pressure tend to cause plants to release more fragrance molecules. In addition, scientists at the Smell and Taste Treatment and Research Foundation in Chicago have found that when some people undergo a rapid pressure change, such as diving into a pool (resulting in higher atmospheric pressure) or going up in an aeroplane (where atmospheric pressure is lower), they experience an 'olfactory window' allowing them to suddenly smell again for a minute or two, even if they haven't been able to smell anything for years.[23]

Some speculate that in weather-sensitive individuals, this same mechanism may kick in, although in a much more subtle form, turning the already exquisitely sensitive nasal receptors into a personal barometer.

Nevertheless the US figures are higher than those generally observed elsewhere and most weather watchers believe that a worldwide incidence of weather sensitivity of around 35 per cent is more realistic. This, of course, is still a substantial number, amounting to one in three.

## Who is most vulnerable?

Average citizens who believe they are weather sensitive are in good company. Many celebrated people including Goethe, Charles Darwin, Benjamin Franklin, Christopher Columbus, Leonardo da Vinci and Samuel Coleridge have attributed their own shifts in mood and behaviour to atmospheric phenomena. Goethe, who was not only weather sensitive but apparently had the ability to predict earthquakes, believed that it was just the 'excellent personalities [who] suffer most from the adverse effects of the atmosphere'. He was, however, quite wrong.

The elderly and the chronically ill tend to suffer more when the weather changes, and women are generally more weather sensitive than men. Children too seem to be much more sensitive than adults. But just as some people are more aware of their own intuition and emotions or the emotional responses of others, some are naturally more in touch with the effects of weather on their general sense of well being.

To understand how some people can predict changing weather, it is necessary to understand weather as a process. The weather we experience today has been building up in our atmosphere for several days. Several studies show that these early changes affect some more than others. It is common, for instance, for weather sensitive individuals to experience symptoms before the weather changes. [24]

Even in those who don't notice or aren't bothered by the way their bodies respond to weather changes, objective testing will show that most do respond to the weather at a biochemical level.[25] Whether we notice it or not, our skin, nose, eyes, nervous system, lungs, membranes and muscles respond to a greater or lesser extent to the weight of the atmosphere, the friction caused by the wind against our skin, the balance of water vapour in the air and changes of temperature, light and electromagnetic impulses.[26]

For some it can be comforting to think in terms of a Gaia philosophy – of ourselves as elemental beings, part of the fabric of the natural environment. Science, far from disproving this idea, has shown that just as individual plants and animals will either thrive or die according to the climate to which they are exposed, individual humans also have complex needs and responses with regard to the weather.

In the most sensitive individuals, practically no part of the body remains unaffected and the most common symptoms of weather sensitivity can include:

- **Bad moods**
- **Depression**
- **Dislike work**
- **Disturbed sleep**
- **Fatigue**
- **Forgetfulness**
- **General malaise (the 'blahs')**
- **Headache (including migraine)**
- **Nervousness**
- **Poor concentration**

- **Rapid or irregular heartbeat**
- **Respiratory difficulties**
- **Rheumatic/arthritic pain**
- **Scar pain (e.g. itching, tingling)**
- **Vertigo**
- **Visual disturbances (flickering)**

These reactions are linked to our endocrine system, the system of glands that regulates the production of hormones in our bodies, and that is affected by pain, stress, seasons and the weather. They may also be linked to our nervous system, which is responsible for the transmission of instructions from the brain to vital organs including the heart, lungs and gut. Human beings are also electromagnetic creatures and as such we are also sensitive to electromagnetic changes in the atmosphere, for instance, those caused by thunderstorms, high winds and approaching warm and cold fronts.

Upsetting the balance in each of these systems brings about noticeable changes in our well being such as joint pain, mood swings, migraine or changes in skin tone. But it can also bring about a host of less obvious changes, for instance in blood composition, immune response and cellular structure.

Our mental health, too, can be affected by weather phenomena. Serious mental conditions such as schizophrenia and manic depression are said to worsen with changes in the weather, and suicide rates are affected too. In particular, heatwaves have been seen to increase tiredness, headaches, insomnia, bad temper and forgetfulness. In hot weather the body produces chemicals that impair judgement and reduce concentration. Thus it can be harder to work productively in over-heated environments, and incidents of road rage and accidents also increase with escalating temperatures.

Hot weather is also linked with higher levels of street violence and attacks, as well as rioting and unrest. Hot, dry winds are said to increase anxiety and aggression. In Germany the accident, crime and suicide rates rise during the föhn, whilst Israeli scientists have found that the hot desert wind, the Sharav, brings a kind of temporary madness.

Winter weather can bring with it a condition known as Seasonal Affective Disorder (SAD). This is a type of clinical depression linked with the lack of sunlight during the winter months. It causes lethargy, sadness, loss of appetite and disturbed sleep. It has been treated successfully in many cases by exposing patients to strong artificial daylight in the form of special 'daylight' bulbs and light boxes that sufferers can use in their homes.

There is also the fact that certain weather conditions are simply stressful. When there is a disruption of a person's routine or daily pattern due to weather changes, their behaviour alters.[27] This type of stress, perhaps combined with other stresses in one's day-to-day life, also sets up a series of biochemical reactions in the body. The growing list of disorders linked to everyday stress – including fatigue, indigestion, infections, irritability, diarrhoea, eczema, headaches, constipation, psoriasis, muscle tension, peptic ulcer, allergies, neck and back pain, irritable bowel, asthma, atherosclerosis, loss of appetite, nutritional deficiencies, high blood pressure, anorexia nervosa, premenstrual symptoms, diabetes, weight changes, sexual problems, arthritis, insomnia, psychological problems and depression – are very similar to those experienced by the weather sensitive.

The concept of weather as a form of stress also seems to square with the scientific finding that stable weather is rarely influential on human health. Instead, it is changeable weather that causes the most problems. Add to this our typically human responses to stress, i.e. self-destructive activities such as smoking, drinking and drug-taking, and you have a body that is severely altered and susceptible to a stress cycle where each stressor or stress-related response feeds into the other.

## Medical tourism

Another way of looking at the effects of weather is to examine the way different climates can benefit human health. The concept of 'climatotherapy' – the idea of recommending different weather conditions as therapy for different illnesses – has fallen out of favour in modern medicine. Yet there was a time when patients with tuberculosis or blood diseases were sent to mountain resorts to benefit from the lower levels of humidity and higher levels of ozone in the air.

At one time, seaside resorts, where the atmospheric pressure is high and the air is rich in sodium and iodine, were considered an optimal place for convalescence and to treat those suffering from exhaustion and from chronic illnesses such as bronchitis and rheumatism. Even today individuals with severe asthma are sometimes recommended to seek drier, sunnier climes in order to improve their condition.

The Dead Sea basin, which lies between Jordan and Israel at the lowest point on earth (about 400 m below sea level), is an internationally recognised centre for climatotherapy. Sojourns to the Dead Sea – where individuals can avail themselves of a combination of sun exposure and sea bathing – have become a popular and effective treatment for psoriasis, atopic dermatitis, vitiligo and other skin diseases.[28]

Its nearest competitor, Safaga in Egypt is a popular Red Sea destination for medical tourism and the treatment of conditions such as arthritis as well as psoriasis.

Higher altitudes too are considered healthful. In Germany, alpine climate therapy is offered on the Predigtstuhl Mountain. At 1,700 m (10,000 feet) above sea level, the dry climate ensures there are no dust mites or mould spores, hardly any pollens, little bacteria, no exhaust fumes and prolonged hours of sunshine. Asthma and many chronic skin conditions as well as certain cardiovascular complaints are believed to improve in such an environment.

Almost in spite of itself, modern medicine has had to admit that for some individuals weather and climate do make a difference to health. What is more, as our scientific understanding of the importance of weather and climate grows, we put a great deal of energy and technical know-how into recreating the perfect climate indoors. Ionisers, air conditioners, heating systems and humidifiers are all part of the arsenal we use to surround ourselves with the perfect, albeit man-made, climate – and all in the name of good health.

# Chapter Two
## Of Seasons and Cycles

All life on earth evolved in an environment ruled by cycles. On a larger, almost unfathomable scale the temperature of the planet changes according to a cycle that spans tens of thousands of years. More easy to grasp are the daily cycles of light and dark, the seasonal cycles that move our weather from warm to cold and back again and even the monthly cycles that trigger a woman's menstruation. Less well publicised, but just as influential to human well being, are what could be called cosmic cycles – relating to the cyclical changes in the geomagnetic forces of the earth as well as those of the sun and the moon.

In his 1947 book *Man, Weather, Sun,*[1] William Petersen introduced the idea of man as a 'cosmic resonator' – in tune not only with the seasonal cycles of the weather but with sun and moon cycles as well. For those who believe that we are all a part of the fabric of nature there is no better phrase to describe our position in the scheme of things.

These days some would argue that our increasing sophistication and the complexity of modern life means that humans are no longer influenced by these cycles. After all, we can turn on electric lights at night to keep the darkness at bay. Women can take birth control pills that replace their natural monthly cycles with a more 'convenient' and regimented man-made one. We have homes, offices and schools that protect us from the extremes and vagaries of seasonal weather. Interventions such as these help support the notion that we are the architects of our environment and are no longer slaves to the cycles of nature.

Our notions of grandeur and power over nature aside, all living things are programmed to survive in harmony with nature's cycles. If man does not also respond to those cycles then he is the only living thing on the planet that does not.

## Rhythm is gonna get you

Biological cycles ranging from minutes to years occur throughout the animal kingdom. These cycles reveal themselves in hibernation, mating behaviour, body temperature, blood pressure and a vast number of other physiological processes.

Every cycle has its own unique rhythm and yet each cycle is also synchronised to many others; our body temperature cycle is, for example, synchronised to our sleep–wake cycles. Some of these rhythms – such as our heart rates and menstrual cycles – are controlled by internal, usually hormonal, factors. These rhythms are called endogenous, because they arise from within the human organism.

A variety of external factors also regulate our body rhythms; our responses to light and dark are a good example. These rhythms are called exogenous, because they arise from outside the human organism. Recent research suggests that these external factors are the greater influence and may even be the true organising force behind our endogenous cycles.

## Chronobiology

Our recognition of the way all life operates in a rhythmic way has been dubbed 'chronobiology' – a relatively new field of scientific endeavour that focuses on the way that both endogenous and exogenous cycles influence biological function.

The cyclical nature of living organisms first came to light in the late 1920s when a French scientist, J.J. Marian, took a heliotropic plant (one that turns towards the light) and placed it in total darkness. Even in the absence of light the plant continued to turn its face towards the 'sun'.

In the 1950s, Franz Halberg, an American biologist at the University of Minnesota, coined the phrases 'chronobiology' and 'circadian' (*circa* = about, and *dian* = day) after his work with mice showed that white blood-cell counts rise and fall in approximately 24-hour cycles. This work spurred other investigations that led Halberg to conclude that all living things respond to the same 24-hour cycle that runs in tandem with the light–dark cycles of the earth. Halberg has since become the acknowledged father of chronobiology and is active in the field even today.

Exogenous triggers are also known as *zeitgebers* (from the German meaning 'time givers'). Common zeitgebers include sunlight, noise, social interaction and conventions such as meal times as well as man-made devices such as alarm clocks.

Halberg's early studies led him to believe that the mechanism for our biological rhythms may have been located in specific brain cells or the adrenal glands.[2] Yet even when these were removed from the animal its innate cycles continued.

Such findings fuelled the conservative view, still held by some scientists, that zeitgebers, especially those arising from the natural world, have very little influence on how our bodies function. Many doctors still believe that our internal clock is the pineal gland, and is largely self-governing or only minimally influenced by environmental cues.

The pineal or epiphysis, is a tiny cone-shaped gland located directly behind the nose. It is sometimes referred to as the third eye and esotericists believe that it is the seat of the human soul. The pineal is exquisitely sensitive to its surroundings and acts as a mediator between man and his environment. In response to light and darkness it secretes the hormone melatonin. Melatonin is widely understood to be responsible for our sleep–wake cycles, but it is also involved in regulating growth and mental stability. Low levels of this important hormone have been implicated in modern health problems such as cancer, hypertension and sexual dysfunction. Malfunctions in the pineal gland have also been linked to epilepsy, schizophrenia and autism.[3]

Halberg did not believe the body's cycles were disconnected from the environment. Indeed, his recent studies confirm that our internal clocks respond to more than just the crude cues of light and dark. The synchronising factor, he believes, lies in the cosmos and in particular in our sun.[4]

Other research confirms this. Not long before his death, Bruno Tarquini, the respected Italian biologist, put forward a similar view. His studies led him to believe that the secretion of melatonin by the pineal gland is organised by two external factors: light during the day and geomagnetic activity at night.[5]

As undoubtedly important as the pineal is, it is increasingly believed that an even smaller area of the brain – in fact a tiny sliver of brain tissue, less than the size of a pinhead – is what truly regulates our body rhythms. Within this sliver lies a biological clock that keeps track of the time of day, and seasons of the year, and organises the responses of our bodies and brains around these.

The small cluster of nerve cells that forms this biological clock is nestled within the hypothalamus and is called the suprachiasmatic nucleus (SCN). According to one theory, the SCN is signalled by messages from the light-detecting retina of the eyes. Having received this message, it can then send a message down the line to other brain centres such as the pineal gland and the hypothalamus to release hormones that control our bodily functions.

So highly developed is the SCN that it has its own dedicated pathway of nerves, the retino-hypothalamic tract (RHT), which is separate from the main nerve bundles carrying visual information to the brain.

We know that the SCN is a biological clock because when it is surgically destroyed in experimental animals, rhythms in sleeping, waking and other daily activities, fade away. Interestingly, an animal, minus its SCN, will run, eat and drink the same total amount over each 24-hour period, but these activities eventually become randomly distributed throughout the day and night.

It is also known that the SCN is easily confused by a lack of zeitgebers. Experiments with humans placed in isolation chambers have shown that in the absence of external cues such as natural daylight, our rhythms are still maintained but tend to deviate from an organised and synchronised 24-hour cycle.

In the absence of natural light our body clocks may lose or gain a little time leading to a desynchronisation of different rhythms. For example, in the absence of sufficient environmental light the sleep–wake and associated rest–activity rhythms may lengthen to a cycle of between 30 and 48 hours, while the temperature rhythm may remain at a period of, say, 25 hours. This desynchronisation can lead to health problems such as hormonal imbalances, sleep disorders and mood disturbances.

## Circadian rhythms

Rhythms that show cyclic changes on a daily (or less) basis are known as circadian rhythms. This cycle moves to an approximately 24-hour rhythm and is probably the best known of the natural cycles.

Circadian rhythms are evident all around us. Plants open and close their petals with the sun. Activities associated with photosynthesis synchronise the release of fragrances and pollens according to the time of day. In the wild, animals follow fairly predictable daily cycles of waking, hunting, eating and resting.

In humans the sleep–wake cycle is the most obvious and extensively studied of the circadian rhythms. At regular intervals each day, the body also tends to become hungry, tired, active, listless or energised. Body temperature, heartbeat, blood pressure and urine flow cycle rhythmically throughout the day. Levels of many hormones, such as ACTH-cortisol, thyroid-stimulating hormone (TSH), and growth hormone (GH), also rise and fall in this relatively predictable, daily rhythmic pattern initiated and governed in part by exposure to sunlight and darkness.

Our daily cycles can work to our advantage. Aerobic exercise is most likely to be of benefit between the hours of 4 and 7 pm, when your heart and lungs are strongest, and your metabolic rate is highest. Taking certain medicines at the right time in the body's cycle may make them more effective (see box page 33).

But most humans tend to ignore the intricacies and demands of their circadian cycles. To live in the modern world many of us work at night and sleep during the day. Even when we are obviously tired we tend to stay up late working or partying. If we have an energy dip in the late afternoon, we tend to meet it with coffee and other stimulants rather than responding to it by resting.

Awake when we should be sleeping and resting when we should be awake it is hardly surprising that our bodies so often develop the symptoms indicative of a system out of sync. Sluggishness and lethargy are the most typical signs, but digestive problems, headaches, depressed mood, poor memory and lack of coordination are also common.

Travelling across time zones can create a very specific type of desynchronisation between our internal clocks and the prevailing light–dark cycles. Known as 'jet lag', its symptoms may also signal that we have become out of sync with the spin of the earth.

Our planet spins from west to east and this is the directional spin in which we evolved. Flying against the spin of the earth – from east to west – generally results in greater disruption of the circadian cycle because it requires the body to delay its release of sleep-inducing melatonin – usually achieved by exposing yourself to lots of bright light and other external cues, even when you are exhausted and need to go to bed. As a rule it takes one day to fully adjust to every one-hour's time zone crossed during a flight.

International travel isn't the only thing that can leave us out of sync. Mistimed conventions such as the time our work or our school starts also have an impact and recent evidence suggests that these may even have a dumbing-down effect on our children.

Our need for sleep varies according to age and teenagers require much more sleep than adults in order to support the tremendous changes and growth that are going on in their bodies. It is believed that teenagers need around 9.5 hours per night to remain healthy yet this amount can be difficult to obtain. One reason is that the release of melatonin in teenagers is delayed (in other words happens later in the night). This is why teenagers commonly feel so alert at night and so sleepy in the morning.

The normal physiology of teenagers plus rising early for school can mean that teenagers accumulate approximately three hours of sleep debt per night.[6] In the US, where schools tend to start much earlier than in other countries, ignoring this fact has implications for students' educational progress. When the school start time was changed from 7.20 to 8.30 am in one US state, the result was higher test scores and fewer behavioural problems.[7]

## Circasemidian rhythms

Within each 24-hour cycle is a shorter 12-hour cycle. A good example of this is the common experience of the 'post-lunch dip' (PLD). While the timing and intensity of the PLD can vary from individual to individual, most of us experience it to some extent each day.[8] PLD can, of course, be caused or made worse by inappropriate food choices (such as high glycaemic index snacks and meals) but there is also evidence that PLD is more than just a nutritional phenomenon.[9] Brain EEG signals emitted during PLD are very similar to those emitted during REM sleep[10] and this has led some to speculate that the PLD is a 12 hourly or circasemidian fluctuation in our sleep–wake cycle.[11]

## Ultradian rhythms

Within each circadian cycle is a much shorter ultradian rhythm, approximately 90 minutes in duration. During the day, we go through many ultradian peaks and troughs[12] which can be experienced as alterations in mood, attentiveness, wakefulness and performance.

Some ultradian rhythms are too subtle for us to notice. Dermatological studies show that skin goes through many changes throughout the day. For instance, the proliferation (growth) of skin cells varies by up to 30-fold over a 24-hour period, being highest at midnight and lowest at noon.[13] In women, blood flow, amino acid content and water loss are roughly 25 per cent higher at night than in the morning or afternoon.[14] Your skin is likely to be twice as oily at noon than it is between 2 and 4 am, is more acidic when you are asleep than when you are awake and has a higher temperature in the evening compared to midday.[15]

Other ultradian rhythms simply get ignored because we live such fast-paced lifestyles. However when we are asleep they are quite pronounced. Throughout the night (or whenever we sleep), our bodies go through an approximately 90-minute cycle involving REM (rapid eye movement, or dream) sleep and non-REM (deep) sleep periods. Disruption of these REM cycles can lead to insomnia, fatigue, loss of concentration and memory and mood disorders.

## Chronomedicine

Understanding our basic rhythms can enhance our health in unexpected ways. The field of chronomedicine explores the interaction between biological rhythms, medicine and drugs.

According to Franz Halberg, 'A long list of drugs can be tolerated without obvious effects, but 12 hours earlier or later the same dose may kill most animals exposed to it.'[16]

Studies show that non-steroidal anti-inflammatory drugs (NSAIDs) may cause significantly less damage to the stomach lining if taken at night rather than in the morning.[17] In one study of individuals with osteoarthritis the incidence of adverse effects was cut in half when NSAIDs were taken at night instead of in the morning[18] and there is some evidence that morning pain can also be controlled by taking NSAIDs at night.[19]

The frequency and intensity of symptoms in arthritic diseases such as rheumatoid arthritis, osteoarthritis, ankylosing spondylitis and gout also exhibit profound circadian rhythms.[20] In people suffering from rheumatoid arthritis the severity of joint pain swelling and stiffness is greater in the morning while osteoarthritis sufferers experience more pain at night.

Likewise, heartburn[21] and ulcers[22] are often worse at night. Research shows that stomach acid secretion is 2–3 times greater between the hours of 10 pm and 2 am than during the day, and that these night-time rises are the result of a circadian rhythm of stomach acid production. Timing heartburn or ulcer medicine to coincide with nightly rises appears to make it more effective.[23]

Growth hormones can successfully stimulate cell division if given at one time of day, but have no effect at other times. Antibiotics, which only kill growing bacteria, work best if they are given with awareness of bacterial growth cycles.

New advances in chronomedicine are occurring all the time with the treatment of asthma and cardiovascular complaints and the timing of breast cancer surgery all showing benefit from synchronisation with circadian rhythms.

## Other cycles

In addition to circadian rhythms the human body is subject to a range of other biological cycles of different frequencies.[24]

There are, for instance, *circaseptan* or weekly rhythms. Some physicians believe that transplant patients tend to have more rejection episodes seven, fourteen and twenty-one days after surgery and that this is reflective of some circaseptan mechanism we don't yet fully understand. Franz Halberg suggests that such responses may be tied in to geomagnetic activity that also follows weekly and even half-weekly (*circasemiseptan*) cycles (see Chapter 3). So did the late Professor Tarquini, whose studies revealed that there were pronounced weekly rhythms in neonatal blood pressure and heart rate.[25] He believed that these were triggered by weekly cycles in the earth's own geomagnetic field and suggested that from the moment of birth humans 'lock into' the pulse of the planet.

*Infradian* cycles are approximately 28–30 days. The most obvious of these is the menstrual cycle with its cyclically fluctuating levels of oestrogen and progesterone. There is now evidence that synchronising breast-cancer screening and breast-cancer surgery to this cycle is important to both the accurate detection and the course of the disease.

Mammograms, we now know, are more accurate when done within the first two weeks after the start of a woman's period.[26] Waiting longer than that means there are increased levels of the hormones oestrogen and progesterone, which cause fluid retention in the breasts and trigger breast cell growth, both of which can obscure readings.

Other evidence suggests that if a woman needs breast cancer surgery, her chances of disease-free survival over ten years are much higher if the surgery is performed during the third week of her menstrual cycle.[27] It's not clear why, but one theory is that during this time progesterone levels are at their highest, which seems to boost the immune system and block cancer cells from spreading.

Another approximately monthly cycle is the *circalunar* cycle. For some women their monthly menstruation also follows this lunar cycle, though opinion is divided on whether the menstrual cycle will naturally, in the absence of modern disruptions, synchronise with the normal lunar cycle (see Chapter 3).

There is also a *circannual* cycle. In the natural world this cycle reveals itself through the seasonal changes in plants and in animal behaviour.

Some seasonal changes – such as Seasonal Affective Disorder (SAD, or winter depression) – appear to be dependeant on day-length. Like other animals, humans secrete melatonin exclusively at night. This secretion is interrupted when exposed to light during the nocturnal period, as well as when daybreak comes. The SCN detects not only light and dark but also changes in the relative length of day and night and makes proportional adjustments in the length of time that melatonin is secreted throughout the night.[28]

Researchers believe that melatonin production increases in winter, leading to an increased urge to sleep and a reduced desire to interact with others.[29] Melatonin levels generally decrease with age, but the elderly do experience seasonal shifts, and in these individuals SAD may worsen with age.[30] Natural or artificial sunlight, especially in the early morning, suppresses melatonin production, thus improving mental health.

Annual cycles are harder and more costly to study than the shorter circadian cycles. However, research in this field has revealed seasonal fluctuations in other hormones.

Bruno Tarquini, for instance, found that compared to healthy women, those with breast cancer appeared to have a malfunction in the annual fluctuation of the hormone prolactin.[31]

In another study, blood samples taken at various times of year from women in the Channel Islands were stored and the women monitored to see which ones developed breast cancer. The healthy women showed an annual cycle in the secretion of prolactin as well as thyroid stimulating hormone (TSH) that was absent in those that went on to develop breast cancer.[32]

## A change of season

For all our apparent ignorance of our natural rhythms, chronobiology research is an amazingly rich field. Current investigations into 24-hour circadian, 90-minute ultradian, month-long infradian and other biological rhythms span across many different fields and have been studied by allergists, animal and plant physiologists, cardiologists, cell and molecular biologists, endocrinologists, environmental scientists, epidemiologists, gastroenterologists, immunologists, neuroscientists, obstetricians, gynaecologists, psychiatrists and toxicologists.

There is some suggestion that our response to seasonal changes in the natural length of the day may have been more robust prior to the Industrial Revolution and that subsequently it has been suppressed by alterations in our physical environment.[33] Some believe that our use of artificial light at night, for example, may have dampened our response to seasonal light changes.[34]

But wider investigation into human cycles show that this isn't necessarily so; a range of common diseases still follow strong seasonal patterns. Many of these are triggered by an inability to adapt to sudden changes of weather. However, they can also be triggered by cyclical changes in our bodies that link into other equally powerful zeitgebers.

## How the body responds

Human behaviour, as evidenced in rates of conception and birth, suicide and mortality, tends to cycle with the seasons.[35] Mood disorders also follow a seasonal trend.[36]

Several studies of human biochemistry and physiology have uncovered measurable seasonal changes in a number of bodily functions. The structure of our blood changes according to the seasons.[37] Blood volume increases in the summer and decreases in the winter.[38] In one UK study, blood pressure was found to be highest in April and May, and lowest in September.[39] In some studies, the basal metabolic rate has been found to be higher in winter than in summer[40] and seasonal variations in body weight are also common.[41]

As temperature drops, the incidence of coronary events and cerebral thrombosis will begin to increase within 24 hours.[42] Our normal physiological response to cold includes a rapid change in many blood components including a decreased total blood volume (leading to reduced oxygen supply), increased fibrinogen levels (promoting coagulation), and increased blood viscosity and cholesterol levels (promoting atherosclerosis).[43] Given the strain that these changes could place on an already weakened heart, it is understandable that coronary events such as angina and heart attack should peak during autumn and winter, and circulatory ailments peak in January and February.[44]

Certain cancers also follow a seasonal pattern. There is a seasonal fluctuation in detection of human papilloma virus and cervical cancer[45] and this translates into a higher incidence of newly diagnosed uterine cervical cancer cases in February.

Our immune response also varies during the year. Most people experience increased circulating leukocytes (infection fighting cells) in August and December, thus experiencing better immunity to infection. However, levels of T-helper cells (also infection fighting cells) are lowest in April and October, increasing vulnerability.[46]

## Being born

Throughout the world, fertility and successful conception appear to be related to both season and weather.

In general, conception rates are lowest in hot weather, although there are some unexplained variations from country to country.[47] For example, seasonal patterns in the United States and Canada differ from those of Northern Europe. One researcher found that in both cool and warm, temperate climates conception is most successful when the average monthly temperature is between 14° to 16° C/57° to 61° F and least successful at 23° C/73.5° F.

In the tropics, however, there is less of a difference, with successful conception occurring when the average monthly temperature is 26° C/79° F, and fewer conceptions occurring when the temperature is 28° C/82° F. In the tropics, high humidity rather than temperature appears to be the factor most likely to interfere with successful conception.[48]

Opinion is divided on whether human interventions such as air conditioning make any difference to these patterns.

In the United States, seasonal patterns in birth rates were once most pronounced in the southern (hottest) states. But after air-conditioning became common, differences in the maximum and minimum temperatures as they related to successful conception became less pronounced.[49]

Some speculate that seasonal diseases may have an influence on conception.[50] In addition, intercourse may be less frequent in uncomfortably hot weather.[51] It may also be that even today women's fecundity may change seasonally.[52] In animals it has been found that sperm counts are depressed by increases in scrotal temperature.[53] Human data agree. A study in Houston, Texas, found that, sperm counts continued to be lowest in the summer despite widespread availability of air-conditioning.[54]

## Birth defects

Certain congenital malformations, for instance of the central nervous and the skeletal systems, appear to follow a seasonal pattern of a winter peak and a summer trough,[55] though this incidence appears to vary from area to area.

In other types of malformations there doesn't appear to be a seasonal link.[56] Some authorities argue that even if there is a seasonal relationship to the occurrence rate of malformation at birth, variations in climate were associated with only a small proportion of these.[57]

Instead, seasonal variations in hormones (possibly linked to climatological or cosmic variables), toxic metals and infectious diseases have all been mooted as the real reason behind seasonal variations in these malformations, suggesting that the influence of the seasons is indirect.

## Schizophrenia

In addition to seasonally varying factors that may damage the central nervous system (such as infection), mothers of schizophrenic patients are more likely to conceive in early summer than are other women. The birth months of people diagnosed as schizophrenic in the US, for example, have been found to follow a seasonal pattern with most occurring in March and April.[58] This seasonal trend tended to be more pronounced in New England and the Midwest than it was in the southern states. Similar patterns have been found throughout Northern Europe, though most scientists are at a loss as to how to explain their significance in relation to the seasons alone.

## Longevity

When you were born may also have an influence on when you die. In one large-scale Russian study involving 101,634 individuals who died in Kiev during the period 1990–2000, longevity was significantly associated with season of birth.[59] Those likely to die youngest were born in April to July, and those that were the longest lived tended to be born at the beginning and end of the year. The difference was small – around two and a half years – but was nevertheless consistent in both men and women. According to the researchers, our rate of ageing may be programmed in response to environmental influences at critical periods of early development.

The fact is that many seasonal rhythms in human health are often hard to explain if one simply looks for obvious external cues such as light and dark or hot and cold. As the next chapter will show, more subtle and seemingly remote cycles may hold the key to our physical as well as our emotional well being.

# Chapter Three

## The Sun and the Moon

Our solar system is an interconnected configuration of planets and stars. Everything that happens within this configuration has some impact elsewhere. Some occurrences, of course, are more dramatic and have more impact on life here on earth than others. With regard to our weather and our well being, two major heavenly bodies, the sun and the moon, are particularly influential.

The sun is a source of incredible heat, light and especially high frequency magnetic energy, and in the same way that humans will respond to electromagnetic waves of our own making as well as those generated by our planet, we respond to those produced by the sun.

The moon's effect is less easy to quantify. Bright and beautiful in the sky, it is also our most erratic cosmic companion, subject to many more cycles than most of us realise. The moon's gravity and its relative close proximity to the earth means that it too can alter the magnetic atmosphere of the earth, and this in turn can influence our weather and our well being considerably.

Both the sun and moon follow their own innate cycles including their proximity to the earth and transition across our sky. As they move through these cycles, so their influence waxes and wanes. While some scientists dispute the idea of cosmic influences on life on earth, believing that such ideas are more at home in the realm of astrology than astronomy, serious study undertaken by space agencies and by reputable independent scientists suggests that in the future such prejudices may need to be reassessed.

## Sun cycles

The scale of the sun and the power contained within it is hard for most of us to comprehend. Our sun has been radiating light and heat for the past four or five billion years. It is so large and so dense that it takes about 50,000,000 years for energy produced at its core to make its way to the surface.

The sun is responsible for our earthbound cycles of light and darkness. Its light-cues trigger the release of the neurotransmitters melatonin (at night) and serotonin (during the day). In combination with the earth's orbit and rotation, the sun is also responsible for the rise and fall of seasonal diseases that depend on changing temperature, the degree of precipitation and the length of the day.

But the sun is also a source of high frequency energy waves – usually related to the appearance of what are called 'sunspots' and their resulting solar flares. While our atmosphere acts like a buffer between the sun's emissions and us, humans and other living organisms are not completely protected from its high frequency energy waves or any of the other astropollution that daily bombards our planet.

As early as 28 BC, astronomers in ancient China recorded systematic observations of the cycles of what looked like small dark patches which move cyclically over the surface of the sun. There are also some early references to sunspots in the writings of Greek philosophers from the fourth century BC.

A sunspot is a dark blotch on the sun's surface that is cooler – at least in a relative sense – than the surrounding area. This dark area, which can be 50,000 miles/80,500 km across, is cooler because it is contained within a strong magnetic field that blocks the transport of heat from the core to the surface of the sun.

The average number of visible sunspots varies over time, increasing and decreasing in a regular cycle of around 11 years. In 1843, an amateur astronomer, Heinrich Schwabe, was the first to systematically study this cycle. The part of the cycle with low sunspot activity is referred to as 'solar minimum', the portion with high activity is known as 'solar maximum'. The year 2000 was the solar maximum for the current solar cycle, though intense activity was observed in 2002. The next solar minimum is due in 2007.

Today we know that the full cycle covers approximately 22 years, with an 11-year cycle of sunspots above the sun's equator followed by an equal cycle below the equator.

The area underneath a sunspot is an area under pressure and occasionally this pressure erupts in a violent explosion known as a solar flare. During a solar flare, highly charged particles – sometimes called 'electrified bullets' – are propelled out into space at speeds of more than five million miles per hour, sending a shower of radiation and ionisation towards the earth.

## Geomagnetic storms

Shifts in the magnetic fields of the sun can also cause what are called corona mass ejections, explosions that throw huge pieces of the sun composed of plasma, energetic particles, x-rays and magnetic energy far into space. Fallout from both types of explosions reaches the earth within one to four days, bombarding us with what are known as geomagnetic storms.

As early as the 19th century, scientists noticed that high levels of activity on the sun, like flares and sunspots, were swiftly followed by strong fluctuations in magnetic instruments on earth.

Ordinarily, the earth's own magnetic field protects us from most geomagnetic storms. But during periods of intense sunspot activity, which coincide with solar flares, the geomagnetic flow from the sun is overpowering. When these storms do eventually reach earth, they have enough clout to disrupt power grids and radio transmissions and even knock out satellite communications.

A few brave scientists are now asking the previously unthinkable question – do these geomagnetic storms also possess enough power to disrupt living organisms?

The answer appears to be yes. It is well known that animals that use magnetism to help them navigate – such as dolphins and whales, butterflies, honeybees and homing pigeons – can be thrown off-course by increases in geomagnetic activity. Even single-celled algae respond to solar flares.[1]

Humans also respond. A recent study showed an increase in intraocular (in the eye) pressure in healthy people during changing and stormy solar activity.[2] Russian research suggests a higher rate of illness and death on unsettled days affected by geomagnetic storms.[3]

Researchers sometimes find themselves at odds with their own remarkable findings in this area. For instance, the authors of one study discovered what they termed 'a remarkable statistical link between sunspot cycles and prevalence of hip fractures in the elderly'. However, as the spectre of scientific credibility raised its ugly head, they concluded that 'the hypothesis of an 11-year cyclic variation of ultraviolet radiation as a cause of hip fractures is untenable'.[4]

Nevertheless, the finding still stands.

Findings such as these often evoke the scorn of conventional scientists and doctors who believe that there is no 'transfer mechanism' between solar activity and humans. Yet even if we don't understand the transfer mechanism now, that doesn't mean it doesn't exist. Indeed it was many years (and many sunburns) before we understood the transfer mechanism between ultraviolet radiation and melanin.

## Weather effects

The sun's effects on earth cycles extend well beyond simply altering the amount of daylight. Studies of its geomagnetic activity have revealed that major geomagnetic fluctuations occur shortly after both equinoxes. These solar cycles are linked to changes in sea level, atmospheric pressure and surface air temperatures in summer, and especially over the oceans in winter.

The extent of Newfoundland's ice cover for the period between 1860 and 1988 fluctuated with solar activity. Waterways such as the Nile, Ohio and Parana (in Buenos Aires, Argentina) rivers rise and fall in concert with solar activity. Ozone levels vary with the long-term solar cycle, as do levels of airborne particles in the stratosphere, and shifts in climate.

When solar detritus hits the earth's magnetosphere it causes a disturbance that can produce heightened, spectacular displays of the aurora borealis and the aurora australis, otherwise known as the northern and southern lights.

The frequency and severity of terrestrial thunderstorms are dictated, at least in part, by solar activity.[5] Data collected from electrosondes (balloons measuring atmospheric electrical currents) over the Antarctic ice caps suggests that solar flares stimulate large surges in the flow of electrical charge from the upper atmosphere to the earth's surface. This unidirectional flow of fair-weather electricity needs to be balanced out, and some scientists believe this explains why auroras in the upper atmosphere are followed by thunderstorms in the lower atmosphere.

## Human effects

There has been relatively little scientific study of human sensitivity to the effects of sunspots and solar flares. A few studies link ultraviolet radiation and changes in magnetic fields to increased metabolism levels or erratic behaviour. During a period of maximum sunspot activity, there may be as many as 200 solar flares in one year – compared with as few as five during a year of minimum activity – and it seems unlikely that this amount of radiation coming our way would not disrupt human sensibilities in some way.

### *Heart disease*

Of all the human systems, the heart is especially sensitive to changes in geomagnetic activity. Heart-attack rates rise and fall with the solar cycles[6] and sudden cardiovascular death rates are higher in periods of high solar activity.[7] Russian studies show that during geomagnetic disturbances the viscosity (thickness) of the blood increases, indeed sometimes doubles, placing additional strain on the heart and raising the risk of heart attack. This risk is greatest within a day of a geomagnetic storm.[8] Heart rate variability – the way that a healthy heart adjusts its beating to different activities such as breathing, resting or walking – is less robust during magnetic storms, another known risk factor for heart attack.[9]

*Infectious diseases*

One of the more interesting findings of sunspot research is that six of the major influenza epidemics, at least as far back as 1917, coincided with the sunspot cycle. All but one of these epidemics involved an antigenic shift – that is to say that the flu virus developed a new coat of protein, which made it resistant to the immunities the population had built up over the years. The only known mechanism through which solar activity might have produced this effect is via penetrating radiation, which is inherently destructive.[10]

According to S.W. Tromp, one-time director of the Biometeorological Research Centre in the Netherlands, solar activity may also lower immunity. Over 30 years of research, using blood data from 730,000 male donors, has led him to conclude that the blood sedimentation rate – which parallels the amount of infection-fighting albumin and gamma globulin in the body – varies with the sunspot cycle. Because of this, the general population's resistance to infection may also follow the lead of the sun.[11]

*Mood and behaviour*

The weather in space also affects the brain and this in turn can produce strange and unpredictable behaviour.[12] There is evidence to show that attempted suicides and hospitalisation due to nervous disorders increases with solar activity.[13] Another four-year study in New York found a link between geomagnetic activity and admissions to two local mental hospitals.[14]

How geomagnetic activity could alter moods and behaviour is not clear, although electroencephalographic (EEG) evidence suggests that during geomagnetically disturbed days the brainwave activity of healthy individuals becomes disturbed and desynchronised[15] and this may show up as altered responses and more impulsive types of behaviour. In already vulnerable individuals this could translate into more mental instability.

While activity is influential, the key element appears to be change (and this is also the case with human responses to weather in general). The greater the variation in solar activity from one day to the next, the more pronounced our responses are.

*Life and death*

Fertility appears to rise along with geomagnetic activity. Not long ago, when the annual number of births from several countries including Australia, Germany, England and Wales, New Zealand, Japan, Switzerland and the USA were analysed, scientists found strong correlations between the 11-year sun cycle and month of birth.[16] The study spanned the years between 1930 and 1984, a period of approximately five sunspot cycles, and showed that 11-year highs in the birth rate could be linked to sunspots, solar flares, magnetic disturbances and changes in temperature.

Other studies show a startling statistical association –

though not necessarily a cause/effect relationship – between a person's life span and the number of sunspots that appeared in the year their mother was born.

A study by credible scientists at Michigan State University, including Franz Halberg, found that if the sunspot activity was at a maximum in its 11-year cycle, children of mothers born at that time would die an average of two to three years sooner than if their mothers had been born during the sunspot minimum.[17]

What could possibly link one's longevity with one's mother's date of birth? The Michigan State scientists noted that when a woman is born all of her eggs are already formed. Later, they will mature and usually be released one at a time in a monthly cycle. However, if sunspot activity is at a maximum near her time of birth, a woman's entire inventory of eggs will be bombarded with higher levels of solar radiation than normal. The ensuing damage might show up as shortened lifetimes for her children.

*Accidents do happen*

The link between accidents and other forms of geomagnetic activity such as thunderstorms has been established (see Chapter 5), but little research exists to link the frequency of industrial accidents with sunspots. S.W. Tromp's research in this area suggested that the more abrupt and dramatic changes in electromagnetic activity within the extremely low frequency range were, the higher the number of traffic violations and industrial accidents.[18]

Perhaps spurred on by such findings, in the early 1970s, a US Atomic Energy Commission-funded project in Albuquerque, New Mexico, produced a report entitled *Intriguing Accident Patterns Plotted Against a Background of Natural*

## War and peace

Some researchers speculate that the effect of sunspots on human behaviour is more profound than we know. According to a theory by A.L. Tchijevsky, the 11-year sunspot cycle is divided into four social periods:

***Period 1***: (approx. three years, minimum sunspot activity): Peace, lack of unity among the masses, election of conservatives, autocratic, minority rule.

***Period 2***: (approx. two years, increasing sunspot activity): Increasing mass excitability, new leaders rise, new ideas and challenges to the elite.

***Period 3***: (approx three years, maximum sunspot activity): Maximum excitability, election of liberals or radicals, mass demonstrations, riots, revolutions, wars and resolution of most pressing demands.

***Period 4***: (approx three years, decreasing sunspot activity): Decrease in excitability, masses become apathetic, seek peace.

Tchijevsky did not believe solar disturbances caused discontent as much as they provided a framework or trigger for it. This trigger could be an increase in geomagnetic activity since geomagnetic storms have been found by other researchers to be associated with increased frequency of accidents, illness, psychiatric hospital admissions, and crimes.

In the same vein, a 1984 article in the journal Cycles noted that periods of international peace also coincided with sunspot cycles. This paper showed that peaceful periods ended 7 out of 11 times within two years prior to sunspot peaks. The probability of this occurring by chance was calculated at less than .008.[19]

*Environmental Features*. In it, the government scientists documented a link between on-the-job accidents of government employees over a period of 20 years with various natural cycles.

This preliminary report found that accidents peaked with the sunspot cycle. They also found that people were more likely to have accidents during the phase of the moon the same as or opposite to that under which they were born.

Shortly after its completion, the report was leaked to *Time* magazine, which lampooned the research under the heading 'Moonstruck Scientists'.[20] Funding for the project was immediately withdrawn and a rich vein of data was consigned to the scientific dustbin of astrology.

## Moon cycles

Unlike the sun, the moon's effect on weather and health appears to be indirect. Indeed its effects may be most easily felt as complementary to those of the sun.

This may be best illustrated by the effect that the moon in different phases has on the waters of the earth. The earth's tides are mainly created by a combination of the force of the moon's orbit, along with the earth's own revolution and gravitational pull from the sun. As the moon passes overhead, its pull drags a bulge of water behind it, with a second bulge created on the opposite side of the earth. The most dramatic tides occur at new and full moons, when the sun, moon and earth fall in line with each other and there is a strong focus to their gravitational force. In contrast, tides vary least at a quarter moon, when the gravitational pull of the moon and sun counteract each other.[21]

The moon, of course, has many cycles. The distance between the moon and the earth varies from about 221,463 miles/356,334 km to 251,968 miles/405,503 km. Thus, sometimes it is in *perigee* (close to the earth) and sometimes it is in *apogee* (furthest from the earth). Its orbital cycle – the time it takes to fully orbit the earth – takes 27.5 days, during which time it moves from perigee to perigee (or apogee to apogee).

The moon also has a *synodic* cycle of 29.53 days, which we measure from our earthbound perspective as the time it takes for the moon to cycle through its four visual phases – from new moon to new moon. Perigee can occur at any phase of the synodic cycle.[22]

Another lunar cycle is the *lunar nodal* cycle, an 18.6-year lunar cycle, which influences weather and other geophysical phenomena. Lunar nodes are points that mark the intersection of the moon's orbit around the earth with the ecliptic – the illusory path of the sun as it revolves around the earth during a year. This lunar cycle is apparent in atmospheric pressure, sea level, precipitation, sea-ice conditions, tidal currents, currents in submarine canyons, sea-surface temperatures, geyser eruptions, volcanic eruptions, earthquakes, thunderstorms, auroral frequency, and biological growth series, including tree-ring widths.

*Syzygy* is the astronomic term for the lunar phases that appear when the centres of the sun, the earth, and the moon lie along a common line. This happens twice during the moon's synodic cycle at the new and full moon. The effect of a syzygy is all the more pronounced when the moon is also in perigee.

The time required for this coincidence to change from one syzygy phase to the other syzygy phase and back again is about 14 months, the duration of the *lunar perigee-syzygy* cycle.

## Health effects

Folk legends are full of the ways in which the full moon brings out the worst in us: more violence, more suicides, more accidents, more aggression. The influence of the moon on behaviour has been called 'The Lunar Effect' or, more unkindly, 'The Transylvania Effect'. The word 'lunacy' is derived from Luna, the Roman goddess of the moon, and from the belief that the power of the moon can cause disorders of the mind.[23] So strong was the belief that the full moon caused mental disorders and strange behaviour that, up until the mid 1800s in the UK, a criminal could still enter a plea of lunacy or 'moon madness' and expect a mitigated sentence.

In some parts of Scandinavia, surgeons refuse to carry out operations during the full moon believing that blood pressure rises and hormone balance changes, making operating more difficult and bleeding harder to stem. Such beliefs are supported by observations of surgeons such as Dr Edson J. Andrews who reported in 1960 that in a study of 1,000 tonsillectomies, 82 per cent of postoperative bleeding crises occurred nearer the full than the new moon – despite the fact that fewer operations were performed at that time.[24]

Other aspects of our behaviour also change with the moon. We eat 8 per cent more and drink 26 per cent less at the full moon relative to new moon periods.[25] We retain more water during the new moon.[26] IVF is likely to be successful if undertaken when the moon is in perigee.[27] [28] Gout attacks peak under both the new and full moons,[29] and we make more doctors' appointments at the full moon.

Modern scientific support for the idea that the moon could influence human behaviour, is often credited to Dr Arnold Leiber, a psychologist at the University of Miami, who analysed 14 years of homicide data in Dade County, (where Miami is), Florida, in 1972.[30] There, admissions to the psychiatric unit at the Jackson Memorial Hospital were higher than normal and the murder rate was three times greater than the year before. Calls to emergency services also rose. Dr Leiber wrote, 'Our results indicated that murders become more frequent with the increase in the moon's gravitational force', and put it down to the fact that, during this time, the moon was only 217,000 miles/349,218 km from the earth – as close as it gets. At this distance he believed the moon exerted considerable gravitational influence.

Others clearly believe this is nonsense. In a review of over 100 studies on lunar effects, a group of professed lunar sceptics concluded that, on the whole, studies have failed to show a reliable and significant correlation between the full moon, or any other phase of the moon, and human health and behaviour.[31]

The researchers blamed media effects, folklore and tradition, misconceptions and personal biases as reasons why so many people believe in the power of the full moon. Such views, however, seem rather harsh and even defensive. There are at least as many studies showing a lunar effect on human health and behaviour as there are those showing none at all.

Crime, crisis incidence, human aggression, human births, and traffic accidents, for instance, are all positively correlated with the phases of the moon.[32] Effects of the full moon on other types of human behaviour such as insanity, alcohol intake, drug overdose, trauma, or the volume of patients in emergency departments are less persuasive.[33] Data on accidents,[34] [35] menstruation,[36] suicides[37] [38] and animal bites[39] [40] shows no clear trends.

Even when there appears to be reasonable evidence of an effect there is always other data to refute it.

*Medical emergencies*

Not long ago, a two-year British study showed a three-percent rise in emergency calls when the moon was full. The researcher, Dr Peter Perkins, concluded that the human body is influenced by the gravitational pull or magnetic forces from the moon which affect the hypothalamus – the part of the brain that regulates sleep cycles, body temperature and hormones.[41]

Another study looked at 7,844 emergency calls to a suicide prevention/crisis call centre over a two-year period. The highest number of total calls was during the new moon. When calls for suicide threats were analysed, there were more calls during the first quarter and new moon.[42]

Similarly, calls to a poison centre monitored over a one-year period showed that unintentional poisonings occurred more often during the full moon, while the number of calls due to intentional poison exposure (suicides/drug abuse) was significantly lower during the full moon and higher during the new moon.[43]

Other studies however don't show this association between the full moon and medical emergencies.[44]

*Violence, aggression and crime*

Several large studies do indicate that we become more agitated and violent around the time of the full moon. In a series involving 11,613 cases of aggravated assault over a five-year period, assaults occurred more often around the full moon.[45] In another study of 34,318 crimes in a one-year period, crimes occurred more frequently during the full moon.[46] Other studies confirm this, with one putting the effect down to human 'tidal waves' caused by the gravitational pull of the moon.[47]

One UK researcher monitored 1,200 inmates in a maximum-security wing at Armley Jail in Leeds for three months and asked them to keep a diary of their mood swings, violent behaviour and aggressive outbursts. There was a marked increase in violent incidents in the day either side of a full moon.[48]

However, other studies show no relationship between the full moon and aggressive behaviour.[49]

*Anxiety, depression and psychosis*

Studies into mood disorders show no clear pattern. When the admission records of 18,495 patients to a psychiatric hospital in an 11-year period were examined in relation to moon phases, admissions for psychosis were highest during the new moon and lowest during the full moon.[50]

A 13-year study of 25,568 psychiatric emergency room visits found that numbers increased near the first quarter moon and a decreased around the new moon and full moon.[51]

In 2000, a Liverpool University study of 100 mentally ill people living in the community found that schizophrenic patients showed significant negative changes in their behaviour at the full moon.[52]

Other data suggests no relationship.[53]

*Birth rates*

Many physicians still agree with the folk wisdom that human births march to a lunar drumbeat. Data from individual birth records for 140,000 live births occurring in New York City in 1968 provides evidence that the timing of the fertility peaks at 3rd quarter of the synodic cycle.[54] The researchers suggested that decreasing illumination immediately after full moon may trigger ovulation – an idea not too far-fetched since there is some connection between light, melatonin and menstrual regularity.

A 1959 study of 500,000 births in New York City, carried out by father-and-son doctors, Walter and Abraham Menaker, found a 1-per-cent increase in births around the full moon.[55] A French review of more than 12 million births agreed, though the effect was smaller.[56] Other studies involving thousands of births also show a 'lunar effect' – but once again there is also evidence of no effect.[57]

## Weather effects

As with the sun, scientists have trouble defining the mechanism through which the moon might affect earthbound weather. Nevertheless, the evidence is intriguing. Floods on the Nile River,[58] as well as coastal storms, follow a lunar pattern[59] and may be at their most devastating when the moon is closest to the earth (perigee) and in line with the sun and the earth (syzygy).[60]

The greatest number of thunderstorms also occurs after either the new or the full moon. Evidence suggests that thunderstorm activity reaches a maximum two days after a full moon and remains high for most of the third quarter.[61] The way that the full moon reflects extra cosmic radiation into our atmosphere may explain this phenomenon since this maximum in thunderstorm activity appears to coincide with maximum levels of cosmic radiation and vice versa.

A century of daily rainfall data from Sydney shows slightly fewer heavy rainfalls at the time of the full moon.[62] Likewise, observations collected from 1,544 weather stations in North America from 1900 to 1949 showed that heavy rain occurs most frequently midway through the 1st and 3rd quarters of the lunar nodal cycle[63] – in other words, rainfall is heaviest about a half week after a new moon and again after a full moon. Correspondingly, a low point in rainfall occurs during the 2nd and 4th quarters with the lowest point of all occurring some three days before a new or full moon.[64] This pattern is not seen everywhere however[65] and different effects may be felt at different latitudes.

In addition, increased rainfall at these two peak times tends to be greater when solar activity was at a minimum, and thus not obscuring the lunar effect. During the years of high solar activity, a powerful solar wind helps to blow the reflective radiation of the moon out of our atmosphere. The difference in effect can be quite dramatic. It has been calculated that the lunar cycle accounts for 65 per cent of the variance in rainfall during years of solar minimum, but only 14 per cent during the year surrounding solar maximum.

The lunar nodal cycle has also been shown to have an impact on our weather.[66] Using sophisticated signal-processing techniques, several researchers have found evidence for 18.6-year variations in atmospheric conditions such as pressure, rainfall and the amount of dust in the atmosphere and weather-related human activities.[67]

## What does it all mean?

So does what goes on in the cosmos affect us? *Yes*. Is the effect as big (especially when it comes to the moon) as some would have you believe? *Maybe* – but this may not be the best question to ask (see below). Is there a great deal more we need to know? *Without doubt*.

Very little research has been conducted into our responses to solar and lunar cycles with regard to hormonal or neurochemical changes in the body. If these cosmic health effects are due to changes in geomagnetic activity, they might well show up as subtle changes in our biochemistry. Already, research has revealed that geomagnetic activity strongly inhibits the production and activity of the enzyme hydroxyindole-O-methyltransferase (HIOMT), which is involved in the production of melatonin.[68] Future research may benefit from looking for changes in other hormones linked with stress or aggression such as serotonin, epinephrine, norepinephrine, testosterone, cortisol, vasopressin, growth hormone, blood pH, 17-OHCS and adrenocrotropic hormone.

But the problems studying the most weather-sensitive individuals also confound research into cosmic resonators.

Science has no method for selecting and studying those who might be more sensitive to the phases of the moon than the average person – any more than it understands why some people are more sensitive to certain drugs. The almost total scepticism towards the idea that any cosmic influence could affect life on earth hasn't helped matters much.

Inconsistent results from studies on moon effect are particularly common. Different scientists study the moon in different ways. Most look at the phases of the moon from new moon to full moon. Some studies define 'full moon' effects as those that occur on the day of the full moon and others consider the effect to be valid if it occurs a few days before and after the full moon.

But this focus on the full moon may be misleading since the syzygy cycle (sun, moon and earth in alignment) occurs both at the new and full moon. Moon phases may also need to be studied in conjunction with other influences, such as the sun, to be truly understood. Studies of the moon at its perigee might have different results from those in apogee. Its effects may be more pronounced in mid-latitudes, due to the relative strength of the magnetic fields in these areas, and is more clearly represented in records during solar minimums when the effect is not obscured by solar activity.

Likewise, the moon's effects on behaviour may have more to do with its effects on the weather. Windiness and low pressure often coincide with the full moon and, as this book will show, this can provoke unusual behaviour.

What may initially seem like a digression in a book about weather is in fact very important. The research brought together in this book suggests that there is a continuum between our earthbound atmosphere and ourselves. Some people may find this too big a picture to grasp. Yet the picture may be bigger still and research into man as a cosmic resonator suggests that what happens in our atmosphere and in our bodies is influenced by an even greater force. That weather is an environmental phenomenon that affects human health is undisputed. Its effects, however, may be stronger or weaker according to where the earth is in relation to other, larger cosmic cycles.

The upshot of all this challenging data may be best summed up with a gentle warning: the human barometer is a sensitive instrument, handle with care.

# Chapter Four

## When the Wind Blows

Wind is a trickster – it is wilful and unpredictable and our dread and fascination with it has a long history. Throughout the centuries the winds that blow continuously around our tiny planet have inspired poets, songwriters, priests, screenwriters and scientists alike to write paeans to their power to both disrupt and restore our lives.[1]

Like many things he did not fully understand, early man rationalised the wind by turning it into a god – or, more commonly, gods. To give something a name is to take control of it, thus to the ancient Greeks, Aeolus was the custodian of the wind gods, loosing breezes, gales or other forces as decreed by the gods. To the Greeks, each specific type of wind under Aeolus's control also had a name: Boreas was the north wind, Euros blew from the east and Notus and Zephyr blew from the south and west respectively.

While Greek wind gods are the most written about, other cultures also deified the wind. Vayu is the Vedic personification of wind. Stribog and Yaponc were the Slavic and Hopi Indian wind gods, respectively. In Japan, where typhoons are common, the wind gods are both colourful and more numerous. Fujin, the Japanese Shinto god of the wind, is seen as a horrific, dark demon in a leopard skin with a bag of winds slung over his shoulder. There is also Haya-Ji, a god of the whirlwind, as well as the terrifying Kami-kaze, the god of wind, storms and vicious cold weather whose name, meaning 'divine wind', was given to suicidal bomber missions faced by Japanese pilots during World War II.

The Greeks and the Romans were the first civilisations to draw pictures of the wind gods on maps. Before compasses were in use, wind icons were how people showed direction on maps. Even as we began to build weather vanes and other devices to enhance our understanding of the wind and its effects on other types of weather, we were still making sacrifices to these wind gods who held so much power over us. Agamemnon, the Mycenaean king who led Greek forces during the Trojan War, sacrificed his daughter Iphigenia to ensure winds strong enough to hasten his army's passage across the sea to Troy.

Wind has long held the power to promote trade (by blowing ships across the ocean) or prevent it (by not blowing). Thus the 'tradewinds' describe gusty winds that carried trade ships across the ocean, whereas to be in the 'doldrums' is to be stuck around the equator where atmospheric pressure is near zero and the winds refuse to blow.

Unbeknown to most of us, our global wind belts also had a role to play in maintaining the 'balance of power' during the cold war. While politicians publicly brayed about the Soviet nuclear threat, secretly they knew that a Soviet nuclear attack on Central Europe was unlikely since it would amount to 'Russia's collective suicide'.[2] Global winds blow primarily from west to east meaning that the fall-out from a nuclear bomb exploded over France's nuclear arsenal in the mountains of Provence would rapidly extend to Kiev and the Ukraine. Dropped in Germany it would poison Moscow within two hours. Logically, whoever controlled the weather had the upper hand and although both sides investigated ways of altering the weather as a means to victory, in 1977 the United Nations produced a treaty signed by 25 nations banning the altering of the weather in times of war.

To scientists specialising in the effect of weather on human health, wind is no less fascinating (though it is somewhat less mysterious) these days and is considered by many to be among the most dynamic and influential of the elements.

## What makes the wind blow?

Wind is air in motion. It develops in response to differences in air pressure caused by the tilt and spin of the earth and the topography of a given area (for instance, the relative balance of land to water).

The tilt and spin of the earth means that the sun heats the air in some regions more than in others. As the air warms, it becomes lighter and less dense, creating low pressure in the area below. Our atmosphere, like our bodies, is always striving towards some form of balance. In a meteorological sense, this means that air tends to flow from areas of higher pressure towards areas of lower pressure until the pressures are equalised. When this happens, the places in between the two pressure zones experience wind. The greater the difference in atmospheric pressure between two areas, the faster the wind speed will be.

To some degree, we have all experienced this process of equalisation in our day-to-day lives. You can feel it when you ascend and descend in an aeroplane. When you open the door to your home, you may notice a burst of air in your face. This is because the pressure inside your house is greater than the pressure outside.

On a global scale, the combination of unequal heating of the earth by sunlight and the earth's rotation means the tropical regions are warmer than the polar regions. This means that there is high pressure at the poles and low pressure at the equator, which accounts for the way in which cold air from the poles cycles towards the equator near the surface and warm air cycles back towards the polar regions at higher levels.

The earth's spin, however, stops this from being a direct route and the flow of air is broken up into three zones between the equator and each pole. These are known as the global wind belts. In each hemisphere, there are three belts. The trade winds blow from north-east and south-east and are found in the subtropic regions. The prevailing westerlies in the mid latitudes blow south-west in the Northern Hemisphere (and north-west in the Southern Hemisphere) and the polar easterlies blow from the east in polar regions.

Geography also plays a part in the global wind belts. Water and land react to heat differently. Land tends to heat and cool more quickly than water. The difference in temperatures means that air is always moving where these two masses meet. This can lead to the formation of tropical storms as often as it does refreshing sea breezes.

## Keeping cool

On a global scale, these wind belts are part of the earth's inbuilt temperature control. Where global winds converge around the equator and in the mid latitudes, warm air rises, low-pressure areas form and rainfall and humidity are likely. Where the global winds diverge (move apart), the cooler air sinks forming high-pressure areas with little chance of rain.

Wind also has a role to play in human temperature control. In this respect winds can be something of a mixed blessing. A gentle breeze on a hot day can bring welcome relief. As the warm air from the land rises, cooler air flows in from the ocean to replace it, creating the gentle breezes we associate with the seaside (late in the evening, the cycle is reversed as land cools more rapidly than water and cooler air from the land blows out towards the relatively warmer sea). Such breezes are so welcome in tropical countries that they have been nicknamed 'doctor winds'.

But on a cold day, wind only makes you colder. Indeed this is the basis of what is known as the wind chill factor – the table that helps us to differentiate between actual temperature and felt temperature on a cold windy day.

Winds assist our own temperature-control mechanisms in two ways. The first is by convection – literally blowing away the thin layer of warm air that always surrounds our bodies. Humans lose about one third of their body heat through convection. Winds can also aid the body in another cooling mechanism, evaporation; on a hot day, a gentle breeze can speed the drying of sweat.

Although windy days can sometimes bring pollen, and with it, hay-fever misery for millions, they can, under the right conditions, help to dilute tree pollens in the air as well as blow away pollution, ozone and other airborne particles.

Exposure to bracing winds can also produce a sense of euphoria, at least in the short term. Such winds are often said to clear the cobwebs from your mind and this too can be considered medicinal. As long as a person can seek shelter before overexposure begins to produce the opposite effect, wind can, in the main, continue to claim its 'doctor' status.

### Ancient wisdom

Without the benefit of modern research methods, the ancient Chinese painted a picture of wind-related illness that is very close to what we know today. Traditional Chinese medicine recognises six different weather factors that influence health. One of these is wind.

According to Chinese philosophy, wind seldom acts alone to create illness. Instead it works in concert with another factor such as cold, heat or damp to cause health havoc. When wind disturbs the human body, it often affects the upper part of the body first, thus head colds, headaches, hay fever and other respiratory complaints, sore throat, runny nose or eyes are all associated with wind. Fevers and mood swings are also a part of the wind-type illness. Because the wind is changeable, the diseases caused by it are as well; they can come on abruptly and leave just as suddenly.

## Winds of change

That wind should be the result of pressure is ironic since pressure, both physically and emotionally, is what many people feel when the wind blows.

In some European countries, the ill effects of the wind are recognised in law. Swiss courts, for example, acknowledge the negative effects of the föhn winds as a mitigating factor in criminal behaviour.[3] A similar warm wind, the Sharav, blows through Israel in spring and late summer and judges are more likely to be lenient in their sentencing for crimes committed during these times.

On both a conscious and unconscious level, wind is stressful. It is disorienting. It moves things about, it takes our seemingly stable world and shamelessly shifts and reorders it. It blows things in your face and up your nose.

This stress response to wind is not just emotional or psychological. If you are out in the wind, you are touched by it. The feel of a gentle breeze on your skin can be soothing. But humans appear to have a surprisingly low threshold to wind. Above a certain speed, the skin begins to transmit warning signals to the brain, making us feel uneasy, perhaps a little anxious and irritable. This is because when a gale is beating against your skin it creates friction. It is drying to your nasal passage and eyes, increasing the concentration of dust, allergens and particulate matter in the air, blowing strange smells your way. Perhaps not surprisingly, the body reacts as if it is under attack.

Some observers believe that when the wind is blowing above 20 mph it triggers some deep primitive instinct in humans that tells us to take shelter, to 'hunker down' somewhere safe until it all blows over. Psychological studies show that wind brings out other primitive instincts in both men and women and each sex responds differently in its presence.[4] Men are restless and tense and tend to lean themselves into the wind, ready to fight whatever is coming. Women tend to be more passive in their response, preferring to seek shelter.

Modern man has not ever really got rid of that dread and awe of the wind. Male or female, young or old, it is likely that a very windy day will bring on a stress reaction – an outpouring of adrenal hormones known as the fight or flight response.

This response is as old as man. When it takes hold, the body automatically draws blood away from the superficial layers of the skin as a defence against attack. There is a general constriction of the peripheral vascular blood vessels including the lining of the membrane of the brain. The adrenal glands begin to pump out stress hormones such as catecholamines (including dopamine, epinephrine [adrenaline] and norepinephrine [noradrenaline]). But this can only go on for so long before the adrenals become depleted, leading to the type of exhaustion that often descends at the end of a windy day.

During these times, levels of other hormones also change. Levels of serotonin, a neurohormone that helps to regulate our moods, may become unbalanced, with a knock-on effect of unbalancing our emotions.

Opinion is divided as to whether winds clear the air or increase the amount of dust, pollen and pollution in it. The truth is that it appears to do both. Winds can clear smog from the cities, and may blow dust and mould spores on to the next town. But it can also break up small particles of irritants and allergens into even smaller ones, making them that much easier to inhale. If the wind is blowing ahead of a cold front, it can also bring allergens from other areas into town before the onset of a storm. For people who suffer from allergies, this too is stressful and debilitating.

The relative temperature and velocity of the wind are things that we can sense consciously. But when the winds blow, several other less obvious things are also happening. For instance, there is a decrease in relative humidity (the air gets drier) and this in turn alters the electrical charge of the air. If the wind is cold it leads to the generation of sferics – low frequency electromagnetic waves that can penetrate

**Valley winds**

While hot dry winds have benefited from the most study, they are not the only winds to influence health. The mistral blows from the highlands of Croatia, Bosnia and Hercegovina, and Montenegro towards the Adriatic Sea. The bora blows out of central and southern France to the Mediterranean Sea. South America's Puna is another cold, dry wind that blows down off high cold plateaus into warmer areas below. It is known to bring discomfort and sometimes dramatic changes in mood and behaviour, caused by the combination of a sudden drop in temperature and changes in atmospheric electricity (see Chapter 5).

buildings and affect brainwave and neurohormone activity (see Chapter 5). If the wind is warm it increases the production of ions – electrically charged particles that also have an effect on brainwaves and neurohormones.[5]

## The ion effect

Air is composed of tiny electrically charged molecules that have both weight and substance. The movement of the air means these molecules bump up against each other causing friction. Many of us experience this as static electricity.

This friction produces ions – atoms or molecules that have gained or lost an electron. As the air molecules jostle against each other their electrons hop from one molecule to another. Those that have lost an electron are called positive ions, while those that have gained an electron are negative ions.

Weather changes are not the only or even the main source of ions in our atmosphere. Most are formed by the radioactivity of the earth's crust and from cosmic radiation. But other natural phenomena including waterfalls, thunderstorms and warm winds can also produce them.

In normal, pollutant-free air over land, there are 1,500 to 4,000 ions/$cm^3$.[6] The normal ratio of positive to negative ions in this same environment is 1.2 to 1.[7]

Ions are unstable and they don't retain their electrical charge for very long, but while they are in the air they are capable of evoking a wide range of responses in a variety of living organisms including bacteria, protozoa, higher plants, insects and animals (including humans).[8]

Laboratory studies have shown that the greatest problems seem to arise when the balance of positive to negative ions in the air becomes altered.[9] A relative overbalance of positive ions – for instance, during the onset of a hot and dry desert wind – can alter the body's biochemistry affecting both physical and mental well being. High levels of positive ions are associated with depression, nausea, insomnia, irritability, lassitude, migraine and asthma attacks, as well as disturbances in the normal function of the thyroid gland.[10] The biochemical processes that lead to these disorders can eventually leave the body exhausted and this, in turn, can lead to an increase in accidents, violent crime and suicides.

Laboratory evidence has shown that volunteers develop a dry throat, husky voice, headache, an itching or obstructed nose and reduced breathing capacity when inhaling air with a high positive ion concentration via the nasal passages.[11]

In contrast, increases in the relative levels of negative ions – at least in the controlled environment of the lab – appear to enhance health in some individuals.[12] The use of negative ion generators (or ionisers) has been found to kill bacteria[13] and significantly reduce levels of microbial air pollution.[14]

Ionisers have also been shown to reduce complaints of headache, nausea and dizziness, and increase alertness.[15] They have been used to treat depression[16] and improve people's ability to cope with stress[17] and may also have a role to play in helping weather-sensitive individuals cope with weather changes.

Nevertheless, there are many differences in the way each of us responds to air ions, indeed some of us do not respond at all.[18] Among the most sensitive are children, elderly and sick people, and persons under stress. Women also seem to respond more to ion depletion in the atmosphere than men and also respond more favourably to an ion-enriched environment.

Scientists are now asking whether an altered ion profile in the uncontrolled environment of nature can have the same effect on our well being. Certainly, when a hot dry wind is blowing, the air is so charged with static electricity that even a handshake can spark a painful electric shock. But can it also produce measurable changes in our body chemistry?

Most of the work in this area has been carried out by ion pioneer Dr Albert Krueger and is based on animal studies, which are not directly applicable to humans. Dr Krueger devoted much of his time to studying the effect of negative ions on mice and then extrapolating his findings to humans.[19]

Nevertheless, Krueger was the first to discover that an excess of positive ions in the air could cause a sudden excessive release of serotonin into the bloodstream[20] – an effect that was subsequently verified by many other scientific investigators.[21] It has also been found that a number of other biochemical systems are also adversely affected (e.g. the production of substances such as catecholamines and other amines, prostaglandins and thyroxines),[22] but changes in serotonin levels are the easiest and fastest to measure, thus these measurements have become a standard when assessing weather/health interactions.

People who suffer from depression will already be familiar with the neurohormone serotonin. While it has been dubbed the 'feel-good hormone' – this is not strictly true. This powerful substance is found in many parts of the body, and, even today, the full extent of serotonin's role in the body is poorly understood. In the brain, it acts as a messenger or 'neuro-transmitter', relaying information about our moods and emotions. Elsewhere, serotonin causes 'smooth muscles' to contract. Vascular (relating to blood vessels), bronchial (relating to the respiratory tract and lungs), and intestinal (relating to the gut) muscles can all be affected by serotonin.

In adequate amounts, serotonin does help to maintain calm and equilibrium. Many modern drugs used to treat depression help the body to maintain serotonin levels and it is thought that in some individuals suffering from depression the body either does not produce enough or does not make good use of available serotonin.

Too much serotonin in your body, however, may cause diverse effects such as migraine, headaches, allergic reactions, irritability, insomnia, hot flushes, tension, swelling, depression, rhinitis, sore throat, bronchial cough, nausea and intestinal spasms. It has also been implicated in Attention Deficit Disorder (ADD).

Krueger's work is also supported by scientists who have spent many years investigating the human effects of the warm Sharav wind in Israel. The Sharav, like most warm winds, is characterised by persistent wind, a rapid rise in temperature and a fall in relative humidity.

### Ill winds

Several winds of ill repute – for example, the föhn (Southern Europe), sirocco (Italy), Santa Ana (United States), khasmin (Near East), and mistral (France) bring with them the ability to induce a multitude of other ills. Such reactions are the stuff of folk legends, but they also have a basis in science. These 'evil winds' are now known to increase the positive ion content of the air.

According to a Swiss meteorological report in 1974[23] the adverse effects associated with positive ion winds include:

**Physical side effects**

Body pains, sick headaches, dizziness, twitching of the eyes, nausea, fatigue, faintness, disorders in saline (salt) budget with fluctuations in electrolyte metabolism (calcium and magnesium; critical for alcoholics), water accumulation, respiratory difficulties, allergies, asthma, heart and circulatory disorders, more heart attacks (approx. 50 per cent higher), low blood pressure, slowing down in reaction time, more sensitivity to pain, inflammations, bleeding embolisms of the lungs, and thrombosis.

**Psychological side effects**

Emotional unbalance, irritation, vital disinclination, compulsion to meditate, exhaustion, apathy, disinclination or listlessness towards work (poor school achievement), insecurity, anxiety, depression (especially after age forty to fifty); increased rate of attempted suicide, larger number of admittance's to clinics in drug cases.

Indeed, alongside Krueger, Professor Felix Sulman, head of the Applied Pharmacology department at Jerusalem University, could be said to be one of the pioneers in the field of weather changes and their effect on human health.[24]

Sulman is one of the founding fathers of the concept of the Serotonin Irritation Syndrome,[25] and believes that an overabundance of serotonin in the human bloodstream lies at the heart of many weather-related ills. Sulman's findings are based largely on his studies of human subjects' reactions to the Sharav. His research shows that almost one third of the Israeli population experience some kind of adverse reaction to this wind; of these, 43 per cent show an unusually high concentration of serotonin in their urine.

Work by other Israeli scientists confirms that the Sharav is a natural positive-ion generator. Nathan Robinson, at the Israel Institute of Technology in Haifa, has found a noticeable increase in the number of positive ions in the air 12 to 36 hours before a Sharav arrives, and that concentrations can reach over 5,000 ions per cubic centimetre when the wind is at its peak. At these levels, the ion concentration coincided with the onset of nervous and physical symptoms in weather-sensitive people and was considered the only meteorological change that could be responsible for the discomfort associated with the Sharav.[26]

A day or two before the onset of a hot dry wind characteristic of the Sharav, sensitive individuals can suffer a range of ills including insomnia, irritability, tension, migraine, amblyopia (dimmed vision), oedema, palpitations, chest pain, respiratory distress, hot flushes, tremor, chills, diarrhoea, increased urination and vertigo – many of the same symptoms associated with serotonin overload. Sulman's studies show that these symptoms abate when treated with negative ions or with serotonin blocking drugs.[27]

That weather sensitivity should improve with negative ion treatment makes sense in the light of other work done in France, Italy, Germany and the USSR on the ionic environment of spas, particularly those situated near waterfalls. The consensus seems to be that the air in many such spas contains a high concentration of ions with a ratio of

## Accentuating the negative

A typical air-conditioned office in the city has only 50 negative ions per millilitre of air (and 150 positive ions) compared with 1,000 negative ions (and 1,200 positive ions) in the same volume of clean, outdoor, country air. So, hiding indoors is no guarantee of protection from ill winds.

Instead, consider other ways to minimise your exposure to positive ions while increasing your exposure to negative ions.

* ***Choose a good quality ioniser*** that produces a minimum concentration of 1,000 ions per cubic centimetre of air, believed to be the minimum necessary to produce a therapeutic response in the body. A well-built ioniser will produce significantly more than this. Your ioniser should also come with a statement from the manufacturer that it produces little or no ozone. A badly designed ioniser can produce high levels of ozone and, with it, nitrous oxide.

* ***Steer clear of synthetic materials*** in your home and office. In air-conditioned buildings, especially those with low humidity, static charges, for instance from electrical equipment, can build up on carpets, furniture, wall fabrics and clothing resulting in an overabundance of positive ions. Opt for natural fabrics where possible.

* ***Take a shower***. Because of its mild re-ionising effect, a shower may help to shake weather-related lethargy.[28]

* ***Open a window***. Ducted air conditioning. Metal ducting attracts charged particles stripping ions out of the air as it passes through the ductwork. Electrostatic filters have a similar effect on ion levels.

* ***Don't smoke and consider using an air purifier***. Smoke and dust particles can act like a sponge, mopping up ions from the air.

* ***Overcrowding***. People remove ions from the air while breathing and each of us carries an innate amount of static electricity.

negative to positive ions being considerably greater than normal and that this contributes, at least in part, to some of the *vis mediatrix* of these resorts.

## Headaches

Writings dating back to the 18th century describe the relationship between weather and migraine. However, according to the Canadian Medical Meteorology Network, it was not until 1981 when scientists Alan Nursall and David Phillips began to scientifically study the effects of weather on migraine in Canada that the connection became official. These scientists discovered that wet, windy and cold weather had a worsening effect on migraine while clear, sunny and dry weather had an ameliorating effect.[29]

Canada's warm Chinook winds have been studied extensively for their ability to affect health, particularly in migraine sufferers. In one study, 43 per cent of those polled cited weather changes as the trigger for their migraine (second only to stress at 62 per cent). In another, when the headache diaries of 13 patients were analysed, Chinook winds increased the probability of headache onset, particularly in those aged over 50.[30] Several studies of Chinook wind conditions, have found that women are more sensitive to weather changes than men.[31]

Nursall speculated that the pathways involved in weather's impact on migraine were connected with serotonin, prostaglandins and various other hormonal agents and this seems to be borne out by other data examining the role of serotonin in migraine.

As with most weather-related disorders, it is the day preceding a change to Chinook conditions, as well as the days that the wind blows, that trigger the worst symptoms in weather-sensitive migraineurs.[32] This is because the weather we experience today is part of a process; the fronts that bring hot winds to a given area often collide a couple of days before the winds arrive.

## Hay fever and other allergies

Although we tend to associate hay fever with springtime (indeed this is usually referred to as 'hay fever season'), hay fever has several 'seasons' throughout the year. It is most common in early spring – when tree pollens are responsible for hay-fever symptoms – and again in late spring/early summer when grass pollens take over. Then it appears again in the late summer/early autumn when ragweed and other weed species release their spores.

When and how much pollen is released from plants is dependent on the weather. Because many plants require exposure to a certain amount of heat before they will release their pollen, a cold or wet spring may delay the release of pollen from trees by weeks. Pollen release can also vary from day to day and even from hour to hour in some places depending on the temperature, humidity, rainfall, wind and sun.

Early mornings are generally the worst time for pollen concentrations because at that time of day plants are busy pumping out pollen and this easily accumulates in the still-stable surface air.

As the air heats up on a sunny morning, warm air rises and

cool air falls adding pollution, dust and pollen to the mix already present. Pollen released in the early morning can travel hundreds of miles from its source, which is one reason why cities can have high pollen counts even if they have very little vegetation.

As the air cools again in the evening, sufferers may feel symptoms increasing. The cool air sinks down to the ground concentrating grass pollen (but not tree pollens which can be released any time) on surfaces.

What this means is that allergens are present in our air even before the wind starts blowing them about. But while wind is the vehicle that can carry pollens into our environment, it may also add to our vulnerability to these allergens through the ion effect.

In the lungs, both positive and negative ions are taken up into the blood stream, where the positive ions stimulate the release of serotonin.[33] Evidence suggests that inhaling air rich in positive ions can reduce breathing capacity by as much as 30 per cent. Less air in means less air out – influencing our lungs' ability to clear themselves of unwelcome irritants.

In addition, in 1972, Albert Krueger observed that negative air ions have an accelerating effect in certain processes linked to respiration.[34] By contrast, a serotonin reaction that can inflame the lining of the nasal passages and inhibit respiration can be triggered by inhalation of positive ions.

Felix Sulman's research concurs. According to his observations, people breathing in positive ions often begin to pant and struggle for breath. It is this struggle for breath that he believes prompts the release of the neurohormone serotonin.

Positive ions also appear to increase the production of histamine in the body, probably because of the way they feed into the human stress cycle. Histamine (the main role of which is to produce gastric juices for digestion of food) is found in many areas of the body. It produces a variety of effects associated with the stress response, including the stimulation of gastric secretion, the constriction of the muscles of the respiratory system, and the dilation of blood vessels. Excess levels have been associated with symptoms very similar to those of excess serotonin, including headaches, allergies, hay fever, nausea and insomnia. Histamine excess is also associated with heat-induced hives,[35] another possible effect of exposure to hot, dry winds.

The stress response may also trigger excess levels of another neurohormone, melatonin. This too may have an effect on certain vulnerable individuals, for instance, those who suffer from nocturnal asthma.

Asthmatics appear to have significantly higher levels of nocturnal melatonin compared to healthy individuals. What is more, melatonin cycles may be delayed in these individuals. In this group, as melatonin increases during the night-time, lung function tends to decrease – something that does not happen in healthy people or non-nocturnal asthmatics.[36]

The various hay-fever seasons also coincide with the ill-winds seasons and the combination may pack a powerful punch. Evidence suggests that in California, where the hot Santa Ana winds (similar to the föhn) blow during fall and winter, more asthmatics end up in hospital emergency departments than at other times of the year.[37] Interestingly, while high winds can distribute irritants and allergens over a wide range, they can also, as previously noted, reduce the size of the particles in the air. Most physicians believe that this should equate with less respiratory illness. But small particles may also slip more easily into the respiratory tract causing allergic-type flare-ups.

**The heart of the matter**

There is an increased incidence of heart attacks during high winds. Romanian researchers have discovered that when the warmer sirocco wind blows across the Dalmatian coasts, circulating levels of thrombocyte aggregates – the fancy name for the proteins that thicken the blood – are higher than during periods of cooler winds or calm weather. In vulnerable individuals this may increase the risk of heart attack.[38] The same team also found that older individuals (those over 50) experienced more erratic heartbeats on these days[39]

## Do you know where your dust has been?

Most of us accept dust as an inescapable part of life. It settles, we wipe it away and it settles again. It's irritating but not necessarily harmful. Yet airborne mould and mildew spores – common right across the three hay fever seasons – can attach themselves to our dust. So can mould spores released from a variety of fungi, from the decay of dead leaves, grass, hay, straw, grains and in the soil. Their growth is encouraged by wet, humid weather and many moulds are not killed by night-time frosts so the 'season' of these can last a long time, in some places all year round.

These days, scientists have discovered that a number of harmful chemicals and micro-organisms are also hitching a ride on airborne particles. Storms in places as distant as China and Africa have generated public attention with dust clouds that travel across oceans to North America, bringing with them living bacteria, fungi, heavy metals and other pollutants. But it's not just pollutants from distant lands we have to worry about. Dust generated in our local areas and containing toxic metals and pesticides may be just as harmful.

Recently, US researchers discovered that burning eyes and lungs, skin rashes and other symptoms of illness associated with winds can be traced to the use of human waste (delicately referred to as biosolids) in agriculture.

The research carried out in Alabama, California, Florida, New Hampshire, Ohio, Ontario, Pennsylvania and Texas, discovered the culprit was the bacterium *Staphylococcus aureus*, commonly found in the lower human colon. In addition, chemicals such as lime, which is added during sludge processing, can irritate the skin and respiratory tract and make people more susceptible to infection.[40]

Though modern treatment can eliminate more than 95 per cent of the pathogens, enough remain in the concentrated sludge that blows into your home to pose a health risk.

Even if it isn't blowing directly into your home, dust from other regions of the world can, when imported on the wind, have an effect on your health. The famous 'red tides' off the Florida coast are a good example.

Red tides are caused by iron-rich dust blowing across the Atlantic from the Sahara desert. When the dust settles in the Gulf of Mexico, it feeds a bacterium known as *Trichodesmium*, which 'fixes' nitrogen in the water, converting it into a form useable by marine life. The result is a bloom of the toxic algae *Karenia brevis*, which chokes the waterways and is dangerous to human health. People swimming in the Gulf at these times can experience skin reactions and respiratory distress from breathing in toxins emitted by *K. brevis*. Eating shellfish poisoned by these red tides can lead to memory problems and even paralysis.[41]

## Changing moods

One of the effects of windy days is an initial 'high' – a feeling that you could take flight or achieve anything. Some individuals experience a short-term increase in productivity – a typical response to the adrenaline rush that follows increased levels of serotonin.

But what goes up must come down. Adrenaline is not as quickly renewed as some other chemicals and once the adrenaline supplies become exhausted, the body must then deal with the effects of excess serotonin. Without adrenaline to balance it out, serotonin becomes a major emotional irritant producing a range of undesirable effects such as anxiety, nervousness, tremors, sweating, dizziness and lightheadedness.

When this happens, the initial high and increased productivity quickly turns into a low and poor performance. This may be why the Israeli Army considers the Sharav, with its high concentration of positive ions, a natural enemy of an efficient fighting force. They even have a term for this reaction called 'Bedouinism', which means the soldiers cease to be alert or effective fighters when the wind is blowing.[42]

In sensitive individuals, the change of moods brought on by positive-ion winds may also be related to excess melatonin. The primary function of this hormone (which is made from the same starting material as serotonin) is to regulate our sleep–wake patterns, though it also helps to enhance immune function.

It is manufactured and secreted in a cyclical way with production normally suppressed during the day – which is why we don't generally feel sleepy then. However, windy days may disrupt this cycle. Too much melatonin as a response to stress, or produced later in the day than is normal, may lead to a drop in body temperature and symptoms of lethargy and sleepiness during the day.

Other researchers say that it is the by-products of melatonin breakdown that may cause mood disorders. Melatonin is usually found only in very small concentrations in the body. In cases of excess, melatonin breakdown products may interact with carbonate and nitrogen dioxide in the body to form substances that resemble other brain-signalling chemicals. Although the extent of the biological function of such chemicals is unknown, laboratory tests show they could have significant depressive effects on behaviour and mood.[43]

This is, however, speculation – though studies into melatonin excess show some remarkable parallels with wind-induced mood changes. There is evidence, for instance, that certain stress-induced conditions such as chronic fatigue and fibromyalgia are rooted in a nocturnal excess of melatonin, rather than a deficiency.[44]

Other stress-linked conditions such as amenorrhoea (absent menstrual periods), anorexia nervosa and depression (especially in women) are also sometimes characterised by high melatonin secretion at night.

## Extreme weather – tornadoes, tropical storms and hurricanes

At the extreme end of the spectrum, violent winds can affect human health in much more direct ways. Tornadoes, tropical storms and hurricanes are generated when warm, light air is propelled rapidly into higher, colder levels by an unstable updraft that can reach over 100 miles per hour. These winds can deprive people of their homes, water supplies and heat and are also capable of causing mass deaths.

Extremely high winds like these are also extremely stressful and the physical effects from tornadoes, tropical storms and hurricanes on those who have been caught up in them, can be measured in the increased adrenal output and eventually exhaustion. Early researchers were adamant that wind storms had a particularly potent effect on the senses, altering our perceptions of taste, smell, hearing and touch.[45] These investigations, mostly by German scientists, have not benefited from modern scrutiny. Neither can scientists agree on whether extreme wind and storm conditions actually improve mental functioning or worsen it.

In one oft-quoted example from 1938, students were taking psychological tests when a hurricane descended upon them.[46] The sky grew dark and the winds reached more than 80 mph and the expectation was that under such stressful conditions the students would perform poorly on their tests. But when these 'hurricane papers' were examined they were found to be exceptional. Massachusetts State College, where the tests were taken, was by no means an exceptional college, standing in the 75th percentile compared to other colleges. But the results of the tests taken during the hurricane put the college in the 95th percentile. The early conclusion was that the storm had somehow made the students more intelligent. However, it is likely that the students simply experienced an adrenaline rush. Other studies have not been able to replicate these findings and it is now generally agreed that the human mind prefers settled conditions.

In any case, quantifying the effects of extremely high winds on human health is difficult. Many of these storms don't last long enough to study their effects. Nor are they easily predictable; indeed the concept of scientists one day 'taming' these deadly storms seems very remote since few can even agree on exactly how they are born.

Tornadoes are giant funnels of circulating air that range from only a few feet to one mile in diameter, and are short in duration (normally only a few minutes long). The wind speed inside a tornado's funnel can exceed 200 miles per hour, enough to turn everyday objects into deadly projectiles. Tornadoes occur all over the world, at every time of the year, but they are most common in the summertime in the Midwestern United States. This region's propensity for tornadoes has earned it the nickname 'Tornado Alley'.

While all thunderstorms are capable of producing tornadoes, detection is a difficult task because not all thunderstorms do generate tornadoes. Weather forecasters can identify the cloud features and conditions that normally precede these storms, and they know where they are most likely to occur. However, predicting the exact time, location, and intensity of tornadoes is still very difficult.

While both tornadoes and hurricanes are spinning columns of air capable of causing great damage, there are important differences between these two powerful storms. Tornadoes are more localised and typically found on land, while hurricanes cover vast areas and draw their power from the warm tropical oceans.

To produce a hurricane, energy from these warm waters is converted into thunderstorms. As these thunderstorms gather together, they can begin to rotate in the same direction, forming a spiral with an eye in its centre. This early, swirl-shaped system is called a tropical depression. As a tropical depression begins to spin and gain power, it can turn into a hurricane.

But just as not all thunderstorms become tornadoes not all tropical depressions become hurricanes. Those with wind speeds of 39–74 miles per hour are categorised as tropical storms. When a storm's wind speed reaches 74 miles per hour and greater, it is categorised as a hurricane. Hurricanes are among the most awesome of nature's spectacles. As one author put it 'A full-fledged hurricane is a vast self-sustaining heat engine 100 times larger than a thunderstorm and 1,000 times more powerful than a tornado. An ordinary summer afternoon thunderstorm has the energy equivalent of 13 Nagasaki-type atomic bombs. Most hurricanes have at least 25,000 times that potential for destruction.'[47]

While hurricanes batter the Americas, similar storms are found in other parts of the world. In the Pacific and Indian oceans and the China Sea, typhoons wreak havoc in coastal areas and on islands that are in the storm's path. On average, these storms are much more destructive than hurricanes. They are also three times more frequent, since they can travel for much longer over warm Pacific waters than they can over the cold waters of the Atlantic.

## Weather phobia

Such is the power of hurricanes and tornadoes and indeed many other types of weathers, that psychologists say they can inspire very real fear and dread in some individuals.

## Afraid of the weather?

| Phobia | Caused by |
|---|---|
| Astraphobia | Thunder and lightning |
| Astrapophobia | |
| Brontophobia | |
| Ceraunophobia | Thunder |
| Keraunophobia | |
| Tonitrophobia | |
| Ancraophobia | Wind |
| Anemophobia | |
| Ombrophobia | Rain |
| Pluviophobia | |
| Lilapsophobia | Tornadoes and hurricanes |
| Antlophobia | Floods |
| Chionophobia | Snow |
| Cryophobia | Ice/frost |
| Pagophobia | Cold, ice and frost |
| Psychrophobia | Cold |
| Homichlophobia | Fog/mist/smoke/steam |
| Nephophobia | Fog |
| Nephelophobia | Clouds |
| Heliophobia | Sun/light |
| Thermophobia | Heat |

Weather phobias are now recognised as affecting a wide range of people throughout the world.

Weather phobics may obsessively watch the forecast for signs of a change in the weather. Reactions to impending severe weather is similar to any other kind of phobia and include physical symptoms, such as laboured breathing, insomnia, nausea and sweaty palms as well as a rising sense of panic.

In the US, thunderstorm phobias rank among the all-time top ten phobias suffered by the population, but the range of known weather phobias goes far beyond a big boom in the sky.

Some weather phobias are easier to identify with and understand than others. It is a given that under the right conditions, exposure to extreme winds and storms can be frightening. But if exposure were the only factor necessary for a phobia to develop, then everyone who experiences the destructive power of weather would be affected.

Storm chasers – individuals who spend their lives following and getting close to tornadoes and hurricanes – report getting an emotional buzz out of this kind of extreme weather. Today, there are even holiday companies that promote storm-chasing adventures. Our differing reactions to extreme weather have led some scientists to suggest that people may differ genetically in ways that make them more or less vulnerable to the emotional effects of trauma – but really nobody can say for sure. Right now, population-based studies show that somewhere in the region of three per cent of adults are storm phobic.[48] What is also known, according to one study at the National Institute of Mental Health in Bethesda, Maryland, is that female storm phobics predominate over males by a ratio of nearly four to one[49] – a gender difference that is documented in numerous studies from around the world.

But in the not too distant future this figure could rise. One of the effects of global warming is likely to be more and more of these extreme types of weather (see Chapter 8). People who have never experienced extreme weather may suddenly find themselves facing frequent and frightening storms. If that is the case, weather phobias may become more and more common.

# Chapter Five

## Stormy Weather

A millennium ago, the law of the Friesians, *Lex Frisonium*, imposed a special fine if an inflicted wound resulted in a weather-sensitive scar. In 1995, after recovering from being stabbed in the back during a tournament, tennis player Monica Seles commented that she could feel an impending change in the weather at the site of her scar. When asked if she had any residual pain from her injury, she noted the scar tingles 'only when rain is coming'.[1] Imagine the field day litigators on her behalf could have if the Friesian law were in force today.

The large body of folk wisdom about how weather affects pain is amply reflected in age-old proverbs such as 'Aches and pain, coming rains', and 'If your corns all ache and itch, the weather fair will make a switch'. These sayings have become part of the fabric of weather lore.

Such direct relationships between weather and well being, especially if drawn by lay people, send medical sceptics howling. Nevertheless, pain syndromes associated with approaching storms are so common and so widely researched that they could be said to be the root of the concept of being 'under the weather'. The Germans even have a word for it – *witterschmerz*, weather pain.

The simple definition of a storm is an abrupt change in the weather caused by the passage of large high- and low-pressure areas swirling and sweeping across a land mass or ocean. Storm activity can be measured scientifically as a drop in barometric pressure, rising temperatures and the formation of huge billowy thunderhead clouds. Often there is a brief period of calm and then pelting rain and gusting wind.

Though such weather often appears to come out of nowhere, in reality storms represent peaks of extreme activity in a cycle of atmospheric changes – from high pressure to low and then back to high again – that is continuously circling our globe.

Sometimes these changes can occur without resulting in rain. They pass so quickly our conscious senses hardly define them as a 'storm'. Yet rain or no rain, and with or without input from the weatherman, the human barometer knows when stormy weather is coming. Most of us are familiar with the feeling; the atmosphere seems to close in on you and you can quite literally feel the weight of the sticky, humid air pressing down on you. Whether the storm reaches a dramatic climax or peters out quietly, there is a tangible sense of release when the barometer goes back up again.

## Today's forecast ... pain

While the results of modern studies can be inconsistent, there is nevertheless a persuasive amount of data to suggest that in certain people – for instance, those with rheumatoid arthritis,[2] osteoarthritis,[3] fibromyalgia,[4] gout,[5] trigeminal neuralgia[6], SLE and Behcet's disease[7], as well as those with missing limbs[8] and scars[9] – the abrupt atmospheric and other meteorological changes associated with approaching storms can mean a worsening of symptoms.

Published research into this phenomenon stretches back to 1877 when the *American Journal of Medical Sciences* included a now famous paper entitled 'The Relation of Pain to Weather'. In it, a Philadelphia neurologist, S. Weir Mitchell, documented the changes in pain perception experienced by a Civil War veteran with a phantom limb when a storm was approaching.[10]

The degree to which weather variables influence pain is not entirely clear.[11] Nevertheless, the sheer number of people suffering from chronic pain syndromes that report being weather sensitive is staggering.

It is estimated that as many as 80 to 90 per cent of patients diagnosed with arthritis report some degree of weather sensitivity.[12] Of course, 'arthritis' is an umbrella term that covers any one of a hundred different chronic pain conditions including the most common – osteoarthritis and the crippling rheumatoid arthritis – as well as gout and fibromyalgia. Thus, it is not surprising that in one study of

### How can 'wet' air be lighter than 'dry' air?

When meteorologists measure barometric pressure they are measuring the weight of the air. This weight can be affected by variables such as temperature as well as the levels of water vapour.

Most people know that water is heavier than air and so find it hard to believe that wet, humid air is lighter, or less dense, than dry air. While it is true that liquid water is heavier, or more dense, than air, the water that makes the air humid isn't liquid. It's water vapour, a gas that is lighter than air. Humid air is actually lighter than dry air at the same temperature and pressure.

Whatever its source – humidity or rain – water vapour causes the density of the air (and the barometric pressure) to decrease. This is why humid air can sometimes have the same effect on chronic pain sufferers as a drop in pressure caused by an oncoming cold front or a fall in temperature.

fibromyalgia patients, 92 per cent of those questioned also reported a weather-sensitive response to pain.[13]

Other chronic pain syndromes,[14] including back pain[15] and headaches,[16] also respond to changing weather. Indeed, studies have shown that more than half of all migraine patients believe that weather is a trigger for their headaches.[17]

## It's all in the mix

Sceptics note that not all studies show a connection between stormy weather and pain[18] and that, sometimes, patient perception and reality are at odds.[19]

This begs the question, why are the results of apparently similar studies so different from each other? In part, it may be that relying on people's memory of pain and on journal-taking is an unreliable way of assessing health. Likewise, studies in office buildings[20] and other artificial environments may not reflect exposure to natural weather conditions.[21]

For instance, in 1958, based on a series of studies of people in skyscrapers, one scientist concluded that people experience more of a pressure change from riding in the elevator of a tall building than is typically experienced during passing weather systems. The conclusion of the study was that if the subjects did not experience a worsening of symptoms going up and down in an elevator, it was unlikely that the small-scale fluctuations in pressure caused by weather phenomena could possibly be influential.[22]

Ingenious as this experiment must have seemed at the time, to get the true picture of weather pain, modern bio-meteorologists believe that it is a more complex mix of weather variables including pressure drop, humidity and/or temperature that causes the greatest stress on the body and most likely leads to increased pain.[23]

This belief appears to be borne out by research. In 1989, German researchers studying patients with lumbar disc prolapses found a link between the number of hospital admissions and decreased temperature and increased humidity.[24] Recent studies of the link between the warm Chinook winds of the Canadian Rockies and migraine[25] (see Chapter 4) also suggest that a drop in atmospheric pressure must be combined with other variables such as a sharp drop or rise in temperature to have a significant effect.

This atmospheric mix is most common just before a storm.

In his 1877 paper, S. Weir Mitchell summed up the fact of pain before the storm succinctly, and in much more evocative language than medics use today:

> **Every storm as it sweeps across the continent consists of a vast rain area, at the center of which is a moving space of greatest barometric depression known as the storm center, along which the storm moves like a bead on a thread. The rain usually precedes this by 550 to 600 miles [885 to 965 km], but before [i.e. beyond] and around the rain lies a belt which may be called the neuralgic margin of the storm, and which precedes the rain by about 150 miles [240 km]. This fact is very deceptive, because the sufferer may be on the far edge of the storm basin of barometric depression, and seeing nothing of the rain, yet has pain due to the storm.**[26]

Again modern research bears this out. In a 1990 Israeli study of weather-sensitive individuals, 80 per cent of those with osteoarthritis and 83 per cent of those with fibromyalgia could predict rain accurately via their symptoms. About three quarters of people with other types of arthritis could also predict rain. In common with the findings of other studies, women were nearly twice as sensitive to weather changes as men (62 versus 37 per cent).[27]

In another 1994 study, the kinds of weather changes that occur four to five days prior to a storm, including raised temperatures and lowered barometric pressure, were significant risk factors for an attack of gouty arthritis.[28]

Interestingly, a change in the weather in warmer climates can be just as influential on the course of chronic pain as a change in a colder climate.[29] The same holds true for different seasons. Decreased temperature and increased humidity can affect pain in both the summer[30] and the winter,[31] putting paid to the notion that weather pain is strictly a cold climate or winter phenomenon.

Any doctor who hasn't had his native curiosity hammered out of him by the experience of medical school and the grind of daily practice must surely be intrigued by such evidence.

## Why does stormy weather affect us?

More than a century after Mitchell wrote about the 'neuralgic margin of the storm', we still don't know a great deal about how or why stormy weather exacerbates pain. Modern science is loath to finance studies into the weather/pain connection, so what knowledge we do have is gleaned from isolated studies throughout the years.

Cynics argue is that there is nothing to understand; that there is no physiological connection between weather changes and increased joint pain. Some note that those suffering from chronic pain syndromes often feel very helpless in the face of their condition. Having something concrete to blame pain on is simply a way of feeling less helpless. Others suggest that the weather connection is more indirect; that during colder weather arthritis sufferers are less inclined to get outdoors and perform normal daily activities, such as gardening, that help keep their joints supple.

Some also believe that it is their psychological state – i.e. depression brought about by dreary weather – that affects arthritis sufferers' perception of pain, making them feel it more acutely. This last point is particularly ironic – suggesting that while sceptics believe weather has no link to physical pain, they have no trouble accepting its unequivocal link to emotional pain.

Whatever the cause – even if it is inexplicable by current scientific research methods – the change in pain symptoms is clearly real for a significant number of people.

A variety of explanations have been put forward to account for this occurrence.[32] There is evidence to suggest that temperature and humidity are linked not only with the subjective experience of arthritic pain but also with an objective measurement of disease activity, the erythrocyte sedimentation rate (ESR) – a biochemical indication of inflammation.[33] Dutch scientists, using a special instrument to measure joint stiffness in 122 people with rheumatoid arthritis and 101 people without arthritis, have also found that people's joints really do become stiffer with increasing humidity, and that this effect is much more pronounced in arthritis sufferers.[34] However, objective measurements like this are thin on the ground.

What is known is that our atmosphere exerts a subtle but continual pushing and pulling effect on our bodies. When pressure or humidity is high, the weight of the atmosphere acts like an invisible, elastic body stocking. It pushes against us and holds us in. When atmospheric pressure decreases and/or humidity increases, it's like having this support garment taken away.

Because tendons, muscles and bones are of differing densities, cold or humid air may cause them to expand or contract in different ways. If there is already swelling, stiffness, inflammation or abnormal mechanics in the joint, as the pressure changes, the unequal expansion and contraction of these tissues may add to these injuries and be experienced as increased pain.[35] The density of scar tissue is also different from that of normal skin. As humidity and temperature change, it will contract and expand at a different rate from the skin around it, resulting in a familiar itching or tingling sensation.

Another theory says that a significant drop in barometric pressure leads to an expansion of air in isolated body cavities and of the fluids in membranes. For instance, individuals whose interests include diving and flying are often more susceptible to a condition known as barodontalgia: tooth sensitivity or pain caused by a change in pressure.[36] Research confirms that this discomfort arises when a change in ambient pressure affects liquids and gases within the body. In the most weather-sensitive individuals, abrupt changes in barometric pressure may likewise affect dental pain either directly or as sinus pain presenting as dental pain.

This transient 'disequilibrium' in body pressure may also sensitise nerve endings and this too may account for increased pain preceding changes in the weather. Finally, seasonal weather patterns do influence mood in some people,[37] and thereby indirectly affect pain perception.

## Mood and behaviour

Changeable weather also seems to be influential on behaviour,[38] especially in children. Years of research in classrooms has shown that children can be extremely sensitive to weather changes, especially when the barometer is heading south.

One of the earliest known studies on the subject was done around 1900 by a public school teacher in Denver, named Edwin Dexter. He studied occurrences of murders and suicides in the Denver area as well as records of corporal punishments in local schools. In a study that lasted 18 years, Dexter determined that there were an above-average number of disciplinary problems when the winds were high and barometric pressure was abnormal.[39]

Other studies[40] and reviews of the evidence[41] concur. In 1963, S.W. Tromp reported that drastic changes in weather conditions were strongly linked to days or periods of restlessness in children.[42] He later noted that this 'unrest' before thunderstorms was probably not due to changes in electric conditions in the atmosphere but to the accompanying thermal stresses. Children's immature thermoregulatory systems can easily be overwhelmed by the double whammy of heat and humidity.[43] Once again it is the combination of weather events that is important.

More recently, a 1993 study in an elementary school in Austria revealed that fatigue, headache, sleeplessness/sleep disturbances, lack of appetite, lack of concentration, dizziness, mood swings/irritability, nervousness, restlessness and impaired performance were all affected by changeable weather.[44] What is more, the irritation level was more than twice as high among girls than boys, especially on unfavourable days.

**Relief from the rain**

Allergy sufferers may also be able to sense approaching rain. Australian researchers have discovered a connection between pollen-induced symptoms and thunderstorms. Their investigations show that rain can rupture the pollen sacs on plants increasing pollen levels by 4- to 12-fold. The pollen is then swept up by the wind and carried along on the air ahead of the storm, leading to an increase in severe asthma attacks just before the thunderstorm erupts.[48]

But not all stormy weather brings allergy misery. Rain can often bring relief to sufferers because it quite literally washes away the pollen from the air. Early morning rains may stop plants from releasing pollens that might otherwise become airborne. But late morning or mid-afternoon rains bring with them a drop in barometric pressure that forces pollen back close to the ground. Symptoms of hay fever and other respiratory disorders may 'predict' oncoming rain at these times

It's not just children that can be affected. In one recent study, scientists asked adult volunteers to perform several mental exercises, from proofreading to memory exercises, while researchers varied the barometric pressure in the room. Small, controlled changes in pressure made the alert volunteers perform better, and the sleepy subjects perform worse. But, when the researchers varied the pressure randomly, mimicking the conditions during stormy weather, all the subjects experienced lapses in concentration.[45]

No one is sure why, but it's thought that such changes in air pressure may cause changes in blood pressure, affecting brain activity.

Laboratory evidence collected from patients during periods of dry heat and intervals of cold rain and wind in Israel[46] concluded that malaise, inactivity, depression and psychological discomfort reported during the cold rainy period could be related to measurable changes in the body's levels of neurohormones and neurotransmitters such as serotonin, thyroxine and steroid, and in amine metabolism which can, in turn, affect feelings of anxiety and lethargy.

The relatively lower levels of oxygen on low-pressure days may also increase the tendency to feel lethargic.[47]

## Extreme weather – thunder and lightning

Stormy weather brings with it significant changes in the electromagnetic activity of the environment. At no time is this more apparent than during that most dramatic phenomenon in the weather's repertoire, the thunderstorm.

That bright flash in the sky followed by a spectacular window-rattling crack sends small children and not a few adults diving under the covers. While we tend to think of such phenomena as rare, our world is constantly under siege from thunder and lightning. At any given moment, there are an estimated 2,000 thunderstorms in progress over the earth's surface. Lightning strikes occur somewhere on earth all the time – about 100 times every second.

Thunderstorms can vary from relatively mild rainstorms to very damaging storms that bring hail and high winds. These dramatic storms form when warm air rises from the earth's surface and moves upwards quickly into the colder levels of the atmosphere. Under the right conditions, this rapid updraft can produce tornadoes, but normally it results in rain, wind, lightning and thunder.

Without lightning, there would be no thunder. Thunder is the noise lightning makes as it travels through the air.

Lightning is produced when updrafts carry moist air high into the atmosphere. As they rise, these water droplets begin to freeze into positively charged ice and snow particles. They also form a cloud. The weight of these frozen particles is such that they cannot be supported on the air indefinitely. As they begin to fall back to earth, they jostle against negatively charged water droplets rising within the cloud. The massive 'spark' that is released is what we experience as lightning.

## Sferics

Our planet is surrounded by electricity. The earth emanates its own electromagnetic fields (EMFs) as well as a multitude of different man-made fields. Radiation from the sun and the moon also add to the EMF profile of our atmosphere. So do changes in the weather.

Thunder and lightning, for example, are largely associated with producing weak electromagnetic fields known as sferics (short for atmospherics).[49] Sferics can also be created by friction, for instance from the action of the wind, and other meteorological events such as the development and movement of warm and cold fronts.[50] They are also powerful enough to penetrate walls – thus the largely indoor lifestyle of modern man is no protection from them.

Meteorologists can use sferics to forecast changes in the weather. Biometeorologists know that their existence can be linked to a whole range of health conditions. This is because the human body is a bioelectric organism and as such is very sensitive to changes in the electrical climate of the atmosphere. When sferics are high – usually one or two days before a storm – pain from migraines, scars and arthritis can also be more intense[51] and violence, traffic accidents suicides and criminal behaviour are also more common.[52]

When atmospheric EMFs rise so do the numbers of ions – electrically charged particles – in the air. While sferics can produce both negative and positive ions, they are mostly associated with the presence of positive ions. When the number of positive ions goes up relative to the number of negative ions, it can trigger the body to release the neurotransmitter serotonin in such quantities that the individual experiences 'serotonin irritation syndrome' (see Chapter 4 for a full explanation) – characterised by anxiety, restlessness, aggression, heightened perception of pain and mood swings.

Positive air ions may also be one reason why thunderstorms can initiate respiratory problems. In two recent studies, researchers at St George's Hospital Medical School in London analysed daily hospital admissions for asthma throughout England and correlated these with the occurrence of thunderstorms within health service regional boundaries. Results showed that days with high sferic (or lightning) activity were associated with a 25 per cent increased risk of asthma admissions[53] and a more than 50-fold increased risk of asthma epidemics[54] in areas that had experienced thunderstorms (see also box page 101).

Positive ions have also been associated with changes in blood pH (with a tendency towards unhealthy acidity) and over-stimulation of the adrenal and thyroid[55] glands and thus eventual exhaustion. They also appear to have a pronounced effect on the circulatory system. They can raise blood pressure and are associated with vascular complaints such as heart attacks,[56] thromboembolism[57] and migraine.[58] In one Russian case study, a healthy man's heart rate was monitored for 50 days and the results correlated with weather conditions over that time. Fluctuations in atmospheric pressure, measured as electromagnetic impulses, were closely linked to fluctuations in his heart rate.[59] It is estimated that around 10 per cent of lightning-related fatalities are cardiovascular in nature, possibly due to the way that lightning discharges affect the electrical circuits in the heart muscle.[60]

As many studies have demonstrated, atmospheric EMFs are able to interact with the pineal gland (the master gland that regulates our daily rhythms, see Chapter 2) and alter its secretion of melatonin.[61] Although this effect is usually seen only with higher frequency EMFs, some scientists speculate that constant exposure to very low frequencies such as those from sferics also have the potential to disturb the human circadian rhythm.[62]

Just who is likely to be most sensitive to the effect of sferics is unclear. However, those whose state of health is deteriorated in some way are more likely to be vulnerable. Likewise, as sferics can alter brainwave activity,[63] people prone to weather sensitivity[64] and those with a high degree of emotional liability (mood swings)[65] may also be more sensitive to such changes.

## Infrasound

Within the broad spectrum of atmospheric EMFs are audible impulses known as infrasound – sonic waves undetectable to the human ear except with special equipment. Infrasound is generated by thunder, earthquakes, large waterfalls, ocean waves, wind, fluctuations in atmospheric pressure and volcanoes. Aircraft, automobiles, rockets and most types of machinery are man-made sources of infrasound.

Infrasound not only possesses electromagnetic power, it is also both the cause of and caused by fluctuations in

## Non-lethal weapons

Experiments with artificially generated infrasound show that it is both physiologically and psychologically disruptive, producing significant negative changes in attention and short-term memory functions, performance rate and mental processing flexibility, as well as changes in the central nervous, heart and respiratory systems.[68]

Those who are not convinced that such effects are real might be interested in the US Army's infrasound weapons programme, or the fact that infrasound is being assessed for its potential usefulness in riot control and other police actions. The use of infrasound weapons is based on the idea that infrasound can be used to incapacitate those exposed to it with nausea and other gastrointestinal disturbances.

One argument against the feasibility of the use of infrasound-based weapons is that infrasound's wavelengths (at the lowest frequencies thousands of miles long) are too long to use with the pinpoint accuracy demanded by today's armies.[69] There is also concern about how to deploy them without also incapacitating the user.[70] Nevertheless, research goes on in this area – and would not receive funding unless there was good evidence that infrasound could cause negative health responses. Though there is little evidence that such 'non-lethal' weapons have progressed much beyond the testing stage, infrasound-generation devices may already have been used for riot control in Northern Ireland.[71]

atmospheric pressure.[66] As with all sferics, infrasound can easily penetrate buildings and affect their inhabitants.[67]

The biological effects of infrasound are similar to those of sferics in general. Its primary effect in humans appears to be irritation.[72] Infrasound waves have been shown to exert significant effects on the biorhythms of the brain[73] and may be associated with sleeplessness.[74] Infrasound may also affect human reaction times. Evidence from the US suggests that there is a measurable relationship between atmospheric infrasound and the rate of automobile accidents[75] and a reduced capacity at work.[76] Natural infrasound may also provoke acute conditions in people suffering from cardiovascular diseases.[77] Some believe that infrasound is at the root of most weather-sensitivity symptoms.[78]

The idea that some unseen force, undetectable to our conscious minds, could affect us so profoundly is unnerving for many. But in our less 'civilised' past there were clear advantages in having a body that is able to resonate with atmospheric electricity. Changes in atmospheric electricity broadcast an early warning bulletin to those who can 'hear' it – a warning signal that within the next day, or within the next few hours, the weather will change, possibly bringing with it a thunderstorm.

That sferics could have this function can be inferred by looking at the animal kingdom. Most animals still have this sensitivity to changeable weather and some show specific changes in their behaviour before the arrival of a thunderstorm. Cats and dogs can become restless when a storm is approaching, cows tend to group together, birds become more active. In response to subtle changes in air pressure that presage a storm, bees will return to their hives and are unlikely to swarm. The saying, 'Swallows high, staying dry;

swallows low, wet will blow' also appears to be indicative of how animal behaviour can predict short-term weather changes. When the weather is fine, the insects that the birds feed on are carried up high on warm thermal currents rising from the ground. As a storm approaches, the insects fly lower towards the ground.

In earlier times, being able to detect this message from the sky was of critical importance for survival, making it possible for humans to forecast and subsequently seek shelter from adverse weather conditions. Today, however, most of us are either completely disconnected or just plain irritated by the bodily symptoms that warn of changeable weather. Instead of seeking tangible shelter, we tend to seek the kind of shelter – painkillers, alcohol, TV and other entertainment – that blots these symptoms out. These things can make the discomfort go away, but they also place an increasingly large wedge between us and our natural environment.

# Chapter Six

## Having a Heatwave

Sunlight is a primary source of energy for all living things. Early man knew this and, whatever your cultural origins, the sun has played a major part in its history and development. The sun god is prominent in every mythology and whether he is called Ra, Apollo, Helios or Surya (or indeed whether *she* is called Amaterasu, Saule or Alinga) is unimportant. In all languages, the sun means power, glory, illumination, vitality, the life force.

Light from our sun travels 93 million miles/150 million km through space to reach us and is now classed as an essential nutrient – as necessary to human life as food, water and air. While recent government decrees about the rise in skin cancer have nearly succeeded in turning the sun into a fearsome enemy, humans instinctively know that they need sunshine to stay healthy.

Adequate exposure to sunlight increases vitamin D production and reduces the risk of rickets in childhood and of osteomalacia (softening of the bones) and fractures in adulthood.[1] It strengthens the heart, aids detoxification, boosts immunity, reduces blood levels of cholesterol (by facilitating its conversion to vitamin D), thus preventing heart disease,[2] and inhibits cancer.[3] It also elevates our moods[4] by helping the body to make good use of circulating neurohormones such as serotonin, and reduces symptoms of asthma and skin disorders such as psoriasis.

In the elderly, pleasantly warm weather can ease painful joints and many midwives believe that brief, sensible exposure to sunlight is an effective way to treat jaundice in the newborn. Indeed, most of us derive some benefit from exposure to sunshine. But when the temperature begins to rise too sharply or becomes too high, it presents a clear and quantifiable risk to human health and stability.

## Maintaining thermal balance

Maintaining a stable body temperature is dependent on a balance between the heat we produce from metabolism and the heat we lose to the environment. As the temperature rises, the brain sends a message to the blood vessels to dilate, allowing more blood to reach the tiny veins near the surface of the skin. This allows the blood to cool before circulating back to the internal organs. We also have several other thermoregulatory mechanisms that aid the exchange of heat at skin level. These are: radiation, conduction, convection and evaporation.

Three of these – radiation, conduction and convection – are considered dry or 'sensible' forms of heat exchange. Through these mechanisms heat can be gained or lost, depending on the environment and the body.

*Radiation* is the process of transferring heat by electromagnetic waves. Humans absorb radiation from the sun and convert it to thermal heat. But like all objects with a temperature above absolute zero, our bodies also give off invisible energy in the form of infrared radiation. This continual give-and-take of radiation helps determine our skin temperature and sense of comfort. Radiation is our primary way of losing heat and, when the air temperature is lower than the body temperature, it accounts for around 65 per cent of cooling.

*Conduction* is the transfer of heat from one object to another. The amount of heat transferred depends on the ability of the object to conduct (or absorb) heat, and the differences in temperature and heat conductivity between objects. Metals have very high heat conductivity, while air has a very low one. In fact, because air is such a poor conductor it can be 'trapped' and used as insulation. Air can, for instance, provide insulation between windowpanes, in fur and feathers, and in between layers of clothes on a cold day. Water on the other hand absorbs heat rapidly. The conductivity of water is 25 times greater than air, which is why a bath or a swim will cool you down faster than sitting in the shade. Conduction via the air accounts for only 2 per cent of our heat loss.

Even though air is a poor conductor, it plays a significant role in a different type of heat loss: *convection*. In the same way that warm air in the wider environment becomes lighter and less dense, air heated by radiation from our bodies also becomes lighter and less dense. As the warm air rises, it is transported away from our bodies and is replaced by cooler air at skin level. Nevertheless, this form of heat exchange only accounts for about 10 – 15 per cent of our heat loss (unless it's a very windy day) and can be a source of heat gain if air temperature is higher than the body temperature.

Our fourth cooling option, *evaporation* – is a wet, or 'insensible' form of heat exchange. Unlike the dry or sensible heat exchanges in which heat can be gained or lost, evaporation can only result in heat loss. Evaporation is the conversion of liquid to a gaseous state at the expense of energy. Humans primarily disperse heat (or energy) by evaporation through the mechanism of sweating.

Sweating can dissipate a tremendous amount of heat. In an environment in which the air temperature is greater than the body temperature, sensible exchange mechanisms cease to function and evaporation becomes the only means of heat dispersal. The drier the air, the more readily water will evaporate. Conversely, if humidity is high the process of evaporation is blocked. Both high humidity and high ambient temperatures are particularly hazardous in terms of keeping cool. Sweating removes heat, by evaporative cooling, five times faster than the body can produce it at rest. But high sweat rates also mean a large loss of fluid and possible dehydration – dangerous when loss of fluid exceeds three per cent of body weight.

The net flow of heat through radiation, conduction, convection and evaporation determines whether one feels warm or cold. If the heat gained is more than the heat lost, one feels warm and vice versa.

## Hot under the collar

Our bodies deal with heat loss and gain in a fairly uniform way. There are, however, huge variations in our individual emotional and psychological responses to heat.

One of the most widely studied, but little publicised, consequences of high summer temperatures is a drastic change in the way we interact with others. Over 100 years ago what was termed the 'thermic law of crime' stated that 'Crimes against the person increase, and those against property decrease, with seasonal and geographic increases in temperature.' Today this is known as the 'temperature-aggression hypothesis' and suggests that high temperatures increase aggression through several (possibly related) psychological and biological processes.[5]

Few of us need such highfalutin theories to confirm what we already know. Our everyday language resounds with imagery that reflects our instinctive understanding of the type of personality changes hot weather can bring. Tempers 'flare' when we fight; we get 'hot under the collar' when frustrated; or we 'do a slow burn' when angered. Heat and anger are inextricably linked in language and myth and so it is in day-to-day life and the impact seems to be greatest on spontaneous acts of aggression.

For instance, in the game of baseball, Major League pitchers are more likely to hit batters with the ball – accidentally or on purpose, it is not at all clear – during games on hotter days than those on cooler days, even when the pitcher's ball control is otherwise deemed to be good.[6]

A more common example that many of us can identify with is the experience of sitting in a hot car in a traffic jam honking a horn in frustration and wanting to murder the person in front of you for not running the red light. Honking your car horn is such a common event that many of us don't even think of it as aggression and yet as the incidence of road rage grows throughout the 'civilised' world, it becomes a useful way to look at the influence of temperature on spontaneous acts of aggression.

In 1976, a US researcher conducted a study in which passing cars were delayed by a driver whose car sat through a green light. The study was conducted when city temperatures were averaging in the mid-80s°F (around 30° C) and involved people whose cars had air conditioning and those whose cars did not. Perhaps not surprisingly, those without air conditioning honked their horns sooner than those with.[7]

In a similar study, researchers in Phoenix, Arizona – where temperatures ranged from 29° to 42° C/84° to 108° F in the spring and summer – followed the behaviour of drivers who were delayed by a car blocking an intersection.[8] These researchers took the study of horn honking to new levels of subtlety by assessing how long it took before the first honk, the number of honks and total time spent honking. Again as expected, there was a significant relationship between temperature and long, loud horn honking. Those without air conditioning were the worst behaved.

Road rage can happen at any time, and whether it is worse everywhere during the summer has not been investigated at

## Going doolally

Our reactions to extreme weather depend to a great extent on our ability to acclimatise. The colonial experience of India and the Far East showed just what could happen when too great a demand is placed on our adaptive mechanisms or when the mechanism fails us in some way. Debilitated by the heat, wives and children were often sent home to England, where they flocked to the comfort of the temperate south coast. Husbands, however, struggled along alone in the heat for years, often drinking themselves to death. In colonial India, the rate of psychiatric breakdown amongst soldiers and civilians stuck in the heat was such that there was a standard process for shipping those who had become mentally unbalanced back home. The process of going mad was called Doolally-Tap, because the transit station for shipping people to England was based at a place called Deolali, near Bombay.

all. However, this might be a worthwhile avenue of exploration. For instance, if cooler in-car temperatures could keep drivers more calm and reduce a worldwide epidemic of road rage, then what is currently a costly luxury might of necessity become a safety feature as essential as airbags.

## Violence on the rise

A favourite subject of early climatological studies, interest in weather and climate effects on aggression and violent crime waned as developments in social science pointed the finger at other determinants, such as social conditions and biological factors. The more social scientists claimed to hold the key to unsociable behaviour, the less climate was believed to contribute to the problem. If weather did play a role, some argued, it was only a minor and indirect one, for instance, by creating a greater number of opportunities for aggression (such as people leaving their homes and shops empty while on holiday, the influx of holidaymakers to tourist destinations and a higher degree of social interaction taking place out of doors).

Culture, of course, also plays its part in the promotion and acceptance of aggressive behaviour and every culture has its own tolerance level of aggression. But given that climate can shape culture, the nagging question is, does climate also shape our beliefs and behaviour?

Available data suggests that it does. When a group of researchers examined evidence from around the world in the 1970s, they found that across many cultures, warmer climates were associated with a greater indulgence of aggressive behaviour, higher homicide rates, less anxiety over widespread acts of aggression in day-to-day life and more human agents of aggression in myths. They also concluded that this link was more strongly related to spontaneous forms of aggression rather than warlike tendencies or revolution.[9]

Many studies have found that seasonal increases in temperature also have a substantial impact on violent acts of aggression.[10] This pattern holds true in both Northern and Southern Hemisphere countries where violent crimes are seen to peak during the summer months and trough during winter months.[11] There is evidence to show that the incidence of family disturbances and assaults is more frequent in summer and least frequent in winter.[12] Other studies show that specific types of violence – for example, wife beating[13] and rape[14] – peak during the summer months.

One large 1987 study looked at the relative frequency of violent and non-violent crimes in the United States (taken from the Uniform Crime Reports, or UCR, which include common crimes such as murder, rape, assault, robbery, burglary, larceny and motor vehicle theft) over the period from 1971 to 1980. Data was then divided into quarter-years that corresponded with the seasons.[15] Temperature data from 240 weather stations was sampled for each year to estimate the differences in hotness between years.

Violent crimes (murder, rape, battery and armed robbery) were more frequent in the summer, with the rise beginning in early spring. What is more, in years when the seasonal temperature was higher than normal, the rate of violent crimes rose higher still. In this study, while the incidence of both violent and non-violent crimes increased with increasing temperature, the rise in rates of violent crimes was much greater than for non-violent crimes.

In an earlier study, the same group of researchers also found a significant link between aggressive crimes (murders, rapes) and temperature, but not between non-aggressive crimes (robbery, arson) and temperature[16] – a finding in keeping with the early thermic law of crime.

Why so much aggression in cities? From an atmospheric perspective, city living creates what are known as heat islands (see Chapter 8) – areas comprised of glass, concrete and tarmac, all of which absorb heat and concentrate it in relatively small spaces. In these urban areas other stresses can build up and eventually boil over into aggressive and even violent behaviour.

Increased heat also changes our physiology, releasing hormones known to increase aggressive tendencies (see page 119). Many city dwellers accept such behaviours as 'typical' of city life, unaware that the very lifestyle that allows them access to theatres, skyscrapers and mass transit also turns them into insensitive, impulsive and sometimes dangerous bullies.

## Death by unnatural causes

The seasonal variation of suicides is a well-documented phenomenon in medical literature. As far back as the late 1880s, it was noted that the incidence of suicide was at its highest during spring or early summer and at its lowest during winter.[17] No one would suggest that weather is the only trigger for suicidal behaviour. Nevertheless, there is enough evidence from studies in both the Northern[18] and Southern Hemisphere countries[19] to suggest that hot weather can be a contributing factor.

Many kinds of meteorological factors are believed to exert an influence on the incidence of suicides including changes in the general dynamics of the atmosphere, variations in sunlight and temperature and falling barometric pressure (of the kind usually associated with storms). Of these, temperature and length of daylight appear to be particularly influential,[20] creating important changes in body chemistry.[21]

Some researchers believe that the magnitude of the effect of weather variables is largely dependent on social factors. For instance, during the warmer months, social intercourse is more intensive than in the cooler months and these may be a source of significant stress.[22] In such circumstances, meteorological factors may act as an intermediary capable of interfering with yearly rhythms in human biological processes[23] and lessening the individual's ability to cope. In vulnerable people, this may lead to higher risk of depressive disorders and, in some, suicidal tendencies.[24]

## Spend, spend, spend

While extreme heat can challenge our adaptive mechanisms and increase mental instability, gentle warmth is a mood enhancer. Among the more eccentric studies that have been done in this area are those that look at our spending habits as they relate to the weather. We spend less time and money in shopping centres as the temperature increases,[25] but oddly our tipping behaviour becomes more generous when the sun is out[26] – perhaps out of gratitude that someone else is doing the donkey work for us during the heat of the day.

However, the most intriguing studies are those done by stockbrokers looking for an edge. Research shows that stock market returns are significantly related to season,[27] and when it is cloudy in New York City, returns tend to be negative.[28] In contrast, one 15-year study found that morning sunshine at the sites of 26 leading stock exchanges around the world – including the New York Stock Exchange – is linked to small but significant positive market returns that day.[29]

One reason for this is that warm (but not hot) weather lifts mood, and that people tend to evaluate future prospects more optimistically when they are in a good mood than when they are in a bad mood.

## How hot weather alters behaviour

Exposure to hot temperatures produces specific changes in our physiology. As with exposures to other types of weather, these effects reflect a complex interplay between a number of body systems.[30]

When hot weather increases blood temperature it has a knock-on effect on our moods and behaviour. In particular, the blood vessels in the cavernous area of the sinuses play a crucial role in cooling the blood that flows into the face and brain. The degree to which blood cools or heats influences the stimulation of the emotional centres of the brain.[31] Cooling the brain triggers the release of certain neurotransmitters, which increase good mood. Heating the brain has the opposite effect.

But it is the stress response to heat – with the accompanying surge in adrenal (fight or flight) hormones such as catecholamines (including dopamine, epinephrine [adrenaline] and norepinephrine [noradrenaline]), glucocorticoids such as cortisol and androgens such as testosterone and dehydroepiandrosterone (DHEA) – that may be more influential. This surge means that we temporarily feel invincible and this certainly squares with the impulsive nature of heat-related crime.

But once depleted, the adrenal hormones that supported us through stress, drop sharply and take us into a deep trough. Vulnerable individuals may interpret the accompanying tiredness and hopelessness in such a way that it triggers an impulsive desire to take their own lives.

Several physiological processes involved in thermoregulation are also involved in emotional regulation. Studies in humans suggest that both an over-response as well as an under-response from the hypothalamus, pituitary and adrenal glands (which form a deeply interconnected system) can produce profound disturbances in social and aggressive behaviour.

When we are hot and/or under stress, we sweat and sweating produces an increase in various corticosteroids from the adrenal cortex.[32] The production of testosterone is influenced by corticosteroids, released when the body sweats. Increases in testosterone levels have been linked to aggression in both men and women.[33] New evidence suggests that some but not all men experience a rise in testosterone levels in the summer due to an increase in the testosterone precursor leutinising hormone (LH).[34] In particular, high testosterone levels have been associated with outwardly directed aggressiveness.[35] Similarly, a relative rise in testosterone in women may mean relative decreases in oestrogen and progesterone – changes associated with female aggression (for instance, during the premenstrual phase).

In the brain, the amygdala, hypothalamus and the hippocampus are also connected via neural pathways and are important centres of thermoregulation. Of these three, the hypothalamus is arguably the most important,[36] since it releases a range of neurotransmitters in response to thermal stress including norepinephrine, epinephrine, serotonin, and acetylcholine.

The amygdala is a kind of emotional switchboard reacting to and assigning significance to incoming stimuli. Its neurons are responsive to changes in heat rate and blood pressure, which can vary with ambient temperature. It can create aggressive responses if the body is under stress.

Most recently, impulsive behaviour in humans has been

linked to an imbalance in serotonin. Studies in Australia suggest that increased exposure to daylight can lead to upsurges in serotonin, provoking depressive or suicidal episodes. According to the researchers, some people may actually be driven to impulsive behaviour such as suicide by anxiety-causing spikes in serotonin levels that occur throughout the in spring and summer, and which follow long winter periods of low serotonin activity.[37]

There is evidence that men who have killed a sexual partner tend to have lower levels of a serotonin breakdown product, 5-HIAA, than the healthy, non-violent males.[38] Serotonin levels are naturally lower in men (by about 20–30 per cent compared to women) and this might explain increased impulsiveness and tendency towards violence in some men in summer. It may also be that some individuals have naturally low serotonin levels – perhaps caused by a genetic glitch that prevents the body from using available serotonin. This too may lead to increased impulsive behaviour,[39] and such individuals may be even more affected by summertime spikes in this neurohormone.

While aggression, violence and suicidal tendencies cannot be explained away simply on the basis of chemical cause and effect, it appears that sunlight and temperature are contributing factors in some individuals. What may show itself as heat-related irritation and impatience in a normal, healthy individual may well turn into acts of aggression, violence and self-destructiveness in others. This could explain, for instance, why those affected by bipolar disorder (manic depression) often experience an exacerbation of their symptoms, in particular mania and aggression, in the spring and summer.[40]

## Summer anomalies – eating disorders and SAD

Several studies have suggested that eating disorders such as bulimia nervosa follow seasonal patterns. The weight of the evidence for bulimia shows that appetite bingeing episodes and weight increase during the winter months and that this type of behaviour may be associated with seasonal affective disorder.[41] But for some individuals, bulimia is worse when the temperature rises.[42] While winter bulimia is associated with measurable changes in levels of specific hormones and neurotransmitters (see Chapter 7), a subgroup of individuals who experience a worsening of symptoms in the summer may be responding to physiological and psychological prompts brought on by sunnier weather.

Women who are insecure about their body shape may, for instance, feel highly threatened by warmer weather that requires them to wear more revealing clothing such as shorts, tank tops and swimsuits. A mild distortion of body image may become much bigger under such circumstances. Logically, it should then follow that if seasonal changes in temperature can bring on such behaviour, some women residing in warmer climates may be more susceptible to bulimic behaviour.

It's not a subject that has received much attention, but one study in 2002 did compare women living in the south-eastern US (Florida) with women living in the north-eastern US (Pennsylvania). Evaluation of the women's attitudes to their bodies, as well as their eating habits showed that those living in the warmer climate of Florida had more concerns about body image, had significantly lower body weight and engaged more frequently in bulimic behaviour such as bingeing, gorging and purging.[43]

The researchers did not believe that the problem of bingeing was weather related, but rather a larger problem in our culture that encourages vulnerable women to behave in such a paranoid way. But evidence on the relationship of summer weather to another summer anomaly may link more directly into bingeing/bulimic behaviour.

Recent psychological reviews reveal the existence of a recurrent summer depression.[44] This has been termed Summer Seasonal Affective Disorder or Summer SAD.

The syndrome was identified by researchers at the National Institute of Mental Health in the US more than a decade ago, and is gaining credence through a series of recent studies in India, China and Australia. According to researchers who first identified seasonal affective disorder (SAD), this 'reverse' SAD affects around 1 per cent of the US population compared to the 5 per cent who suffer from conventional winter SAD.

The symptoms of summer SAD – insomnia, poor appetite, weight loss, agitation and anxiety – commonly strike in April and May when seasonal temperatures begin to rise. While winter SAD is believed to be the result of decreased production of melatonin (see Chapter 7) the causes of summer SAD are less clear.

It has been suggested that the same factors that trigger summer depression might also be risk factors for suicide,[45] though studies that have looked at either the prevalence of SAD among suicide victims or the incidence of suicides among patients with SAD are thin on the ground.

Most early findings suggest summer depression is related to seasonal shifts in temperature, though some researchers think that longer daylight hours may play a role. In one brain-imaging study of nine patients with a summer SAD, abnormalities were noted that suggested a problem with glucose metabolism in the brain,[46] but more data is needed.

Current treatments include air conditioning, avoidance of excessive daylight and antidepressants (which can lower body temperature). In extreme cases, sufferers are recommended to move to a more favourable climate, though manipulations of environmental stresses, such as heat and humidity, have not yet been proven to be as effective in summer SAD as light therapy is for winter SAD.[47]

## Skin damage

Summer sunshine generates a great deal of ultraviolet radiation. The three primary effects of this type of radiation on human health are damage to the skin, eyes and immune system. A fourth problem, photosensitivity, is less well understood but increasingly common.

Prolonged exposure to ultraviolet (UV) radiation causes 'sunburn', a reddening and sometimes blistering, of human skin. An individual's sensitivity to radiation depends on their skin pigmentation, with fairer-skinned people being more susceptible to sunburn.

As a result of frequent UV-light exposure and sunburn, the skin experiences what is called photo-ageing, with its characteristic wrinkles, dryness, altered pigmentation and a loss of elasticity in the skin. Photo-ageing is simply a premature ageing of skin cells caused by the production of free radicals brought on by excessive UV exposure. This kind of damage is most prevalent in hot climates but also in polluted areas where decreased ozone levels mean greater exposure to UV radiation.

Photosensitivity is another type of skin reaction to sun. People with photosensitivity break out in a rash when exposed to sunlight; how much exposure it takes to cause this reaction varies from person to person. Several conditions, such as erythropoietic protoporphyria and polymorphous light eruption, share the common symptom of hypersensitivity to sunlight. People taking certain prescription drugs (sulfonamides, tetracycline and thiazide diuretics) or herbs (St John's wort, for example) and those with systemic lupus erythematosus are generally more photosensitive than others.

Symptoms may include a pink or red skin rash with blotchy blisters, scaly patches or raised spots on areas directly exposed to the sun. The affected area may itch or burn and the rash may last for several days. In some people, the reaction to sunlight gradually becomes less with subsequent exposures.

When there is a profuse sweating and the body is not exposed to air, red pustules the size of mustard grains appear on the body, especially on the chest, back and abdomen. This is known as prickly heat because it feels as prickly as thorns. Prickly heat can be avoided by using lightweight, loose cotton clothing and by bathing at least twice a day with cool water.

## Skin cancer

The two most common forms of skin cancer are basal cell carcinoma and squamous cell carcinoma. Carcinomas are cancers that originate in the cells that cover an organ. These cancers are also referred to as non-melanoma cancers.

Basal cell carcinoma appears as small, fleshy bumps, 80 per cent of which are found on the head and neck. Basal cell carcinoma rarely spreads, but if left untreated it can penetrate to and damage the bone.

Squamous cell carcinoma appears as red, scaly patches or nodules. While less common than basal cell carcinoma it can metastasise, or spread, to other organs. Both basal and squamous cell carcinomas are common among Caucasians, especially those that freckle easily (usually those with red or blond hair), but are rare among dark-skinned people.

With increased ultraviolet radiation reaching the earth, rates of actinic keratosis, which are thick, scaly growths usually found on the face, hands, forearms, the 'V' of the neck or places commonly exposed to the sun are increasing. Actinic keratosis is premalignant, and is a risk factor for squamous cell carcinoma.

Another form of skin cancer is melanoma, which begins in the melanocytes (cells that produce pigment). Melanoma forms when melanocytes become malignant and commonly begins as moles or benign skin growths that are clusters of melanocytes and supportive tissues.

Melanoma is the most aggressive form of skin cancer, and it is known to spread and be fatal to humans. If cancerous cells reach the lymph nodes, the cancer can be carried to other parts of the body such as the lungs, liver, brain or other organs. Melanoma is thought to have its roots in childhood sunburns as well as the increased amount of time people spend in the sun today.

While melanoma is believed to be related to the amount of solar radiation, it may also have some relationship to female hormones because, especially under the age of 55, the incidence of skin melanoma is higher in women than in men.[48]

However, the frequency of exposure and the amount of skin exposed to solar radiation differ according to culture and the male/female ratios for melanoma differ with climate. Thus, it could be argued that there may be a climate effect on hormone balance that makes women more vulnerable.

## Immune system damage

Some scientists believe that UV light causes skin cancer through the combined effect of suppression of the immune system and damage to DNA. As UV rays are absorbed by the body, there is a decreased immune response. On the one hand, this reaction is healthy because there is no excessive swelling and damage to the skin as a result of sun exposure. The drawback of decreased immune response is that when diseases do attack the body, a significant forceful immune reaction is needed.

The theory goes that excessive exposure to sunlight triggers immunosuppression that prevents the immune system from recognising and destroying malignant cancers. If you already have a premalignant growth on your skin, more sun exposure may prove to be the crucial factor as to whether it develops into a malignant growth or not.

The degree of sensitivity to immunosuppression by UV radiation varies among individuals. In a small study conducted at the University of Miami School of Medicine, sun exposure followed by dermal application of a substance known to provoke a hypersensitivity (immune) reaction resulted in a vigorous immune response in the majority of exposed individuals. But around 40 per cent of volunteers appeared not to react at all suggesting some degree of immunosuppression.[49] But why some of us should be so affected by sun exposure while others are not is unknown.

## Eye damage

Damage to the human eye from sun exposure is most commonly in the form of photokeratitis and cataracts.

Also known as 'snow blindness', photokeratitis is the ocular equivalent of sunburn. It is very common in skiers and other outdoor enthusiasts and occurs after an acute, short-term exposure to UV light during which the cornea of the eye is sunburnt. Photokeratitis is characterised by reddening and inflammation of the eye, eyelid twitching, photophobia (avoidance of light) and blurred vision.

Cataracts – a gradual loss in the transparency of the lens of the eye – are the leading cause of blindness. There are several types of cataracts, categorised according to their location in the eye. *Cortical cataracts* form on the outer layer of the eye, known as the cortex. *Nuclear cataracts* form on the inner layer of the eye, known as the nucleus. Behind the interface of the lens, *posterior subcapsular cataracts* form. The fourth type of cataract is known as *mixed*, and is a combination of any two of the other types of cataracts. Mixed and cortical are the most common types of cataracts. Research has directly linked cataract formation to exposure to UV radiation, and the prevalence of cataract formation after age 30 is doubling each decade, possibly due to increased ozone depletion.

## Don't blame the sun

Each of these radiation effects is preventable and to a large extent we are blaming the sun for disorders that are largely a result of our own bad habits.

Smoking, for example, can triple your risk of developing skin cancer.[50] There is evidence that diet also plays a major part in our bodies' reactions to the sun. Studies show that skin cancer victims have more polyunsaturated oils in their skin cells. Most of the polyunsaturated oils in our diet come from sunflower, safflower and other vegetable oils. Once consumed, these oils work their way to the skin surface where the effect of sunlight oxidises them quickly and creates free radicals – unstable molecules that damage the cells' DNA leading to the deregulation we call cancer. (Likewise the cellular membranes in our eyes can contain high concentrations of polyunsaturated fatty acids, and this too can lead to cellular damage.)

So-called 'healthy' polyunsaturated fats have recently been linked to immune system suppression of the type that leads to cancer,[51] particularly in the breast.[52] Since 1974, the over-consumption of this type of fat has been blamed for the alarming increase in malignant melanoma across the globe from Scandinavia to Australia.[53]

In 1987, researchers analysed 100 melanoma patients and the same number of healthy individuals for levels of fatty acids in subcutaneous fatty tissue.[54] Compared to the healthy people, melanoma patients' fatty tissue had much higher levels of omega 6 polyunsaturated fatty acids. The researchers reasoned that increased consumption of dietary oils such as sunflower oil had a contributory effect in the development of melanoma. But it's not just people in sunny climates who are at risk. One recent study of more than 50,000 men and women with cutaneous malignant melanoma in Norway also linked excessive polyunsaturate consumption with a significantly increased risk of melanoma in women.[55]

## Covering up and blocking out

Sunburn and the immune system damage it can cause is preventable by staying out of the sun during peak hours between 11 am and 2 pm. It can also be prevented by wearing appropriate clothing when out in the sun. Likewise, UV eye damage is largely preventable by wearing sunglasses.

Clothing provides useful protection and most summer clothes provide an SPF of more than 10 – thus specially designed clothing (reputed to block UV rays) is not only expensive, it's unnecessary. An average weight t-shirt provides an SPF of 7.56. According to one report, measurements of over 5,000 fabrics in an Australian laboratory showed that 97 per cent of fabrics fell into this category (and more than 85 per cent of the fabric samples had an SPF of 20 or higher).[57] In a European study, 67 per cent of fabrics tested had an SPF of 15 or more with more than 70 per cent of wool polyester and fabric blends showing an SPF of 30 or more (compared to less than 30 per cent of the cotton, linen and viscose fabrics tested). Fabrics dyed black, navy blue, white, green or beige provided the highest SPF.[58]

Suncream is a different matter. In spite of the millions pumped into government campaigns, covering up with suncreams is fraught with uncertainties. Used properly, a sunscreen will prevent sunburn – but the evidence of their effectiveness against most skin cancers is pretty thin.[59] While sunscreen use may reduce the risk of squamous cell carcinoma,[60] its effect on the more serious basal cell carcinoma and the more deadly malignant melanoma is much less clear. Indeed, recent studies have shown a higher rate of melanoma among men who regularly use sunscreens and a higher rate of basal cell carcinoma among women using sunscreens.[61]

The latest thinking is that sunscreens and blocks may actually increase the risk of melanoma,[62] though some scientists dispute this.[63] It is not yet known whether it is the sunscreens themselves or the false sense of security they bring – encouraging fair-skinned individuals to stay out longer in the heat of the day[64] – that is responsible for the association. Some scientists believe that sunscreens actually promote the formation of free radicals[65] and that it is through this mechanism that they increase the likelihood of skin cancer.

## Consequences of avoiding the light

Out-of-proportion worries about skin cancer and increasingly sedentary indoor lifestyles mean that many otherwise healthy, adequately nourished individuals are not getting enough sunlight. Unfortunately, modern urban populations spend large amounts of time, including leisure time, indoors and this has led to a growing deficiency in vitamin D.[66] The problem is especially acute in northern latitudes where sunlight is at a premium anyway. But even in sunny Australia many individuals may have insufficient sun exposure to allow adequate synthesis of vitamin D in the skin.[67]

The fact is we need sun. Most of the body's vitamin D supply – between 75 and 90 per cent – is generated by the skin's exposure to UVB rays.[68] As little as 15 minutes exposure several times a week can trigger the synthesis of vitamin D, a substance that is more like a hormone than a vitamin. Without adequate sunlight exposure, there is a risk of vitamin D deficiency disorders such as rickets, osteomalacia and osteoporosis.

But recent research suggests that vitamin D and sunlight may be important to human health in ways that are unrelated to their effects on bone.[69] It may have an early role to play in the development of schizophrenia (due to effects of low prenatal vitamin D on the developing brain).[70] Many body tissues have vitamin D receptors, and the active form of vitamin D (colecalciferol) is also believed to have an important protective role in the initiation, development and spread of various types of tumours.

Specifically, vitamin D deficiency may trigger development of the prostate, breast and colon cancers[71] as well as a number of immune disorders such as type 1 diabetes mellitus and multiple sclerosis (MS).[72]

The prevalence of childhood type 1 diabetes increases with increasing latitude and decreases with adequate vitamin D in infancy.[73] There is also a gradient of increasing MS with increasing latitude. This is as true for high northern latitudes as it is for high southern latitudes.[74] For instance, in Australia there is a stronger correlation between annual UV levels and MS than there is for the incidence of malignant melanoma.[75] A large body of evidence suggests that UV stimulates the production of vitamin D, thought to be protective against MS. Other explanations include the idea that UV radiation acts on certain parts of the immune system to suppress autoimmune activity.[76]

It is, of course, very easy to draw a direct relationship between sun on the skin and the development of skin cancer. Less obvious (and so virtually ignored) is the pathway by which sun prevents other types of cancer. Yet, according to recent review, adequate sunlight exposure may well prevent death from a range of reproductive and digestive cancers.[77]

Using a sunscreen drastically lowers the cutaneous production of vitamin D3.[78] Because of this, Dr Gordon Ainsleigh in California believes that the use of sunscreens actually causes more cancer deaths than it prevents. His work has suggested that the 17-per-cent increase in breast cancer observed between 1991 and 1992 in the US may be the result of the pervasive use of sunscreens over the past decade.[79]

In the US, the number of deaths from skin cancer is believed to be in the region of 10,000 per year. However, some studies suggest that each year the number of premature deaths from cancers associated with too little sun exposure (breast, prostate, colon) is double or more that figure.

In one study, deaths from low UVB exposure amounted to 21,700 per year.[80] Dr Ainsleigh puts the figure higher suggesting that in the US some 30,000 cancer deaths could be prevented each year if people would adopt a regimen of regular, moderate sun exposure.[81]

Clearly the problem is in our perception of relative risks. Most people do not appreciate how wide the gap is between adequate sun exposure necessary for health and the dose required to produce skin cancer or damage to the eyes. For example, a person with a moderately fair complexion living in Boston need only to expose 6 to 10 per cent of their body surface (face, hands and arms) to midday sunlight in spring, summer and autumn for five minute, two or three times a week to maintain sufficient vitamin D levels.[82] In contrast, the same person living at a more temperate zone in Western Australia would need to be receiving 14 hours of sun exposure per week every week for life to be at significant risk of basal cell carcinoma (the most deadly form of skin cancer).[83]

## Extreme weather – heatwaves

Even in the normally adaptive human body, there is a limit beyond which excessive heat can lead to problems like syncope, cramp, exhaustion, heatstroke and, in some vulnerable individuals, death. The heatwave that swept through Europe in the summer of 2003 left more than 10,000 people dead.[84] Most of these were elderly persons. It was the hottest summer on record in Europe and proved how unprepared we still are for abrupt and dramatic changes in our normal weather patterns.

Extremes in heat can cause a range of different problems ranging from the mild to the life threatening.

*Heat oedema* is probably the mildest form of heat-related illness and occurs when swelling – in the fingers or ankles, for instance – occurs in hot weather. It usually disappears on its own if the person rests with legs elevated or after the person becomes fully acclimatised.

*Heat cramps* are painful spasms in the arms, legs or abdomen, usually seen in adults who sweat profusely (such as those taking exercise in extreme heat) but who take fluids without an adequate amount of salt. Cramps occur in heavily worked muscles especially when the person is relaxing. Cold showers may also trigger cramps. Heat cramps – which may be a warning sign of impending heat exhaustion – can be prevented by increasing salt intake.

*Heat syncope* involves dizziness and/or fainting after prolonged standing in the heat. It's more likely in those who have exercised without a cooling-down period afterwards, in people who are dehydrated and the unacclimatised. The sufferer recovers if allowed to rest flat on their back in a cool room. The use of lightly salted liquids will also help.

*Heat exhaustion* occurs when a person experiences heavy sweating in a hot, humid environment. The core body temperature may rise above 38° C/100° F. Symptoms include profuse sweating, malaise, headache, dizziness, nausea, vomiting, vertigo, chills, muscle weakness and visual disturbances. Resting in a cool area for 2–3 hours with plenty of fluids usually promotes recovery.

*Heatstroke* is a very dangerous illness. It can be caused by hot, humid weather or may be induced by some medicines (see page 137). With heatstroke the body's core temperature may rise to 40 ° C/104° F, a point at which damage to tissues in several organ systems can occur. Emergency management of heatstroke includes removal from direct sunlight, removal of clothing, wetting the body or immersing it in cooler (not cold) water and fanning. Severe cases may require hospital treatment and as many as 10 per cent of those who experience heatstroke will die.[85]

## Killer heat

When unexpected hot weather hits an otherwise temperate region the daily numbers of deaths increase.[86] While our bodies make a number of physiological adjustments to extreme heat over time,[87] when the rise in temperature is fast, there is no time for the body to acclimatise. A heatwave in London in 1995, for example, caused a 15-per-cent increase in death from all causes.[88]

## What raises the risk of heat-related illness?

Heat stress has been extensively studied, more so than cold-related stress and other weather influences. The consensus among the research[89] is that there are several factors that raise an individual's risk of succumbing to heat-related illness. These include:

- **Dehydration** – Impairs cardiovascular and thermoregulatory function by decreasing skin blood flow and sweating rate, leading to a decreased ability to disperse heat.
- **Extremes of age** – the elderly are unable to effectively disperse heat due to age-related autonomic dysfunction, neurological and cardiovascular disease, use of multiple drugs that affect heat dispersal, increased obesity, decreased cutaneous blood flow, poor physical conditioning and reduced sweat production. Children are at risk because they have higher metabolisms, a decreased ability to sweat and slower acclimatisation mechanisms
- **Heart disease** – Patients with chronic cardiovascular disease may be unable to compensate for the increased stress from heat exposure.

Infants and the elderly are at increased risk in a heatwave.[90] Older people are more at risk because the main stress is on the circulatory system and many older people have heart and vascular diseases. Because of circulatory problems, their thermoregulatory systems are not as efficient as those of younger people.[91] Excess mortality in infants less than 24 hours old has been noted; this is probably because their thermoregulatory systems are not yet functioning adequately to counteract the heat.

Nevertheless, the majority of excess deaths that occur during heatwaves occur among those with pre-existing illness – for instance, cardiovascular, cerebrovascular, and respiratory diseases – which have been made worse by the heat.[92]

Our cardiovascular system in particular is challenged by

• **Other illnesses** - Including diabetes, hyperthyroidism, eating disorders, cystic fibrosis.

• **Obesity** – Some (but not all) obese individuals have decreased vascularity (fewer fine veins) in the tissues close to the skin; this can inhibit heat dispersal by decreasing blood flow close to the skin.

• **Skin diseases** – Scleroderma, cystic fibrosis, eczema, psoriasis and burns decrease sweating ability.

• **Clothing** – Inappropriate outerwear can prevent heat dispersal.

• **Poverty** – People in the lower socio-economic status generally were found to have the largest increases in deaths because they tend to live in the inner cities, frequently in high-density, possibly inadequately ventilated, housing. A high percentage of very poor people may also work in jobs that expose them to extreme heat.

Certain medications also raise the risk of heat-related illness. These include:

• **Anticholinergics** – impair the sweating response;

• **Diuretics** – lead to decreased blood volume cardiac output, resulting in less sweating;

• **Phenothiazines** – antipsychotics deplete central stores of dopamine and interfere with thermoregulatory centre of hypothalamus;

• **Heart drugs** – beta blockers, calcium channel blockers and alpha agonists decrease cardiovascular response to heat, reduce peripheral blood flow and ability to sweat;

• **Alcohol and drug abuse** – inhibits production of antidiuretic hormone leading to relative dehydration; they can blunt stimulation to leave a hot area and seek a cooler environment.

hot weather. In the heat, our blood vessels dilate sending more blood and fluid to the upper layers of the skin and extremities. This has the immediate effect of increasing heat loss by altering the electrical responses of the skin (the same responses that betray our emotions in a polygraph test). Generally speaking this is a good thing and it helps to keep us cool. But, as the blood vessels dilate, blood pressure drops and the heart has to work harder to pump blood around the body. In the unacclimatised, those with already weak hearts or those engaged in strenuous activity in hot weather this can translate into increased risk of heart failure.

Death rates can increase markedly as a result of heatwaves, and the highest death rates are usually 1–2 days after the temperature reaches its peak.[93] Some speculate that most of those who die during heatwaves would have died soon anyway. But if this were the case, there would be a compensating drop in mortality rates in the weeks and months after a heatwave. This is not always the case,[94] though the data is not at all conclusive.[95]

There are, however, significant differences in death rates between the first heatwave of the year and the second. Death rates in the second tend to be little different from normal suggesting that acclimatisation is protective.

Nevertheless, the most influential factor appears to be whether or not you are a city dweller.[96] Daytime temperature differences between city and countryside can be small. But at night, country air cools faster than city air because soil and grass have lower heat conductivity and storage capacity than brick and concrete surfaces. Large urban structures such as high-rise flats also lose heat more slowly than single-family dwellings. As a result, more heatstroke victims can be found in upper levels of multi-storey buildings than in suburban homes.[97]

While some studies have suggested that widespread use of air conditioning has reduced the effects of heatwaves, the actual protective effect of air conditioning is relatively small.[98] Air conditioning may lessen heat stress temporarily. But the bigger picture of air-conditioning use suggests that it may interfere with our adaptive mechanisms to heat,[99] making people more vulnerable to heat if, for instance, a power outage cuts off air conditioning in their homes and offices or they venture outside into the non air-conditioned environment.

# Chapter Seven

## Winter Chills

If anthropological theory is to be believed, humans owe their robust existence to extremely cold weather. The last ice age was the catalyst that drove us down from the trees, forced us to migrate to the plains and eventually separated our lineage from that of the gorilla and the chimpanzee. Life on the plains was a continual challenge, but it nevertheless honed all our survival skills and helped define humans as the adaptable species we are today.[1]

Our instinctive desire to get out of the cold may hearken back to some collective experience of how lethal cold exposure can be. Statistics tell us that cold weather is responsible for significantly more illness and excess deaths than hot weather. The study of the human body's reaction to cold under every kind of natural and contrived condition reflects our continual desire to find the optimal way of protecting ourselves from this threat. To this end, many modern technologies have been utilised to invent weatherproof clothing, better home, office and school insulation and winterproof transportation.

It's not just our bodies that respond to cold. Our minds and emotions do too. In the natural world, autumn and winter represent the end of a cycle of activity that spans from new growth to flowering to fruiting and finally to 'death'. Autumn is harvest time, when we begin to stock up for the long dark winter months. Our bodies know this even if our minds don't. Body heat is generated in part by digestion and this is the time of year when people feel hungrier and less satisfied by a meal that would have left them full in the summer. In the autumn, our energy intake, especially from carbohydrates, can increase by around 222 calories per day.[2] Modern research suggests that autumn (not winter as many assume) is the time we are most likely to put on weight – possibly as a kind of thermal insurance against the coming cold. Autumn's blustery, changeable weather can also bring on the same kinds of health problems – headaches and allergies – that windy days at any time of year can.

Winter, in the natural world, means slumber and hibernation. Plants and other animals understand this though humans have long since lost touch with this particular rhythm – at least on a conscious level. Even though we fight it in our minds, our bodies do appear to initiate their own mini shutdown, which we can experience as lethargy and a low mood and which suggests a natural inclination towards some form of winter hibernation.

Apart from digestion, the body employs two other mechanisms to keep warm and conserve energy. Because the internal organs need to be kept at a constant temperature, when a cold snap hits, the signal goes out to restrict blood flow to the external shell – the superficial layers of tissue and the skin – and thus reduce heat loss via radiation and convection. This withdrawal of blood from the extremities is why your fingers and toes, nose and ears are often the first things to feel the cold.

If you still haven't managed to find shelter and warmth, your body will initiate another mechanism – shivering. What we experience as shivering is actually a series of involuntary muscle contractions designed to produce heat. This shivering mechanism tends to kick in much faster if the body is undernourished, stressed out, fatigued or experiencing anxiety or pain.

However, both mechanisms are protective only over the short term. The body cannot go on forever withdrawing blood from the extremities, nor can it continue to shiver indefinitely. These are signals that you need to find shelter and get warm. If you don't, the likely result is frostbite, hypothermia and even death (see page 160).

## Adaptation to cold

While extremes in hot weather (heatwaves) still present the highest risk of premature death, overall the number of deaths on an average winter day is still 15 per cent higher than on an average summer day.[3]

Nevertheless, the impact of cold on human well being is highly variable. It can be responsible for direct causes of death such as hypothermia, but it is also a factor in a number of indirect causes of ill health and death such as influenza and pneumonia, as well as falls, accidents, carbon monoxide poisoning and house fires.[4]

Although logic would dictate that the colder the climate the greater the risk of cold-related illness and death, this is not necessarily so. Once again, what you are used to is important. In one study comparing winter mortality rates for 13 cities in different climates around the US, there was a significantly greater death rate when cold weather hit otherwise warm regions in the south, whereas northern regions, where the population was used to the cold, suffered much less.[5] In

Minneapolis, Minnesota, for example, the researchers found no excess deaths even when the temperature dropped to -40° C/-40° F; but in Atlanta, Georgia, death rates increased dramatically when the temperature hovered around 0° C/32° F.[6]

Our ability to adapt quickly to a sudden drop in temperature, then, is our best defence – underscored by the fact that the most crucial time for illness and death appears to be during the first harsh cold spell of the season. The longer the temperature remains low, the more we tend to acclimatise. Members of the armed forces, modern adventurers and professional sportsmen and women often use modern thinking about acclimatisation to their advantage, exposing themselves to simulated extremes in temperature before travelling in order to strengthen their adaptive mechanisms. There is evidence, for instance, that men who bathed in 15° C/59° F water for one half-hour each day in the nine days prior to a trip to the Arctic, showed fewer signs of cold-induced stress than non-treated men.[7]

On the other hand, our ability to adapt to winter chills can be made less effective if we heat our homes, schools and offices too high during this time. While indoor heating (combined with better hygiene) has resulted in somewhat lower rates of death from respiratory disorders in the winter, it has not significantly altered the death rate from coronary events.[8] A heated building means that the impact of going out in the cold is all the more stressful and shocking for the heart. In midwinter, the difference between indoor temperature and outdoor temperature can sometimes be as great as 10°–15° C/18°–27° F. Under such circumstances, our adaptive mechanisms may become inefficient, our respiratory tract may respond with spasm to the sudden inhalation of cold dry air and our immune response may be dampened leading to eventual illness.

## Heart disease

Of particular interest to biometerologists are the apparent seasonal increases – in both the Northern[9] and the Southern Hemispheres[10] – in the incidence of cardiovascular-related illness and death.[11]

During the winter, we consult our doctors more often about cardiovascular complaints[12] – and with good reason. The number of coronary events such as angina and heart attacks peak during autumn and winter,[13] and circulatory ailments peak in midwinter (January and February).[14] The rates of sudden death and stroke at this time are particularly startling;[15] some estimates suggest these are at least 35 per cent greater in the winter than in the summer.

This increase has been linked to changes in temperature,[16] exposure to daylight[17] and the way that these and other weather-related effects change the structure of our blood.[18] While each of these things is a contributing factor, the risk of heart failure in the cold begins with our thermoregulatory responses.

When the weather is cold, the heart has to labour much harder to help keep the body temperature normal. Vasoconstriction – the narrowing of blood vessels that helps to draw blood away from the body shell – also raises blood pressure and this can also place unaccustomed strain on the heart. An already weakened heart is much more at risk than a healthy heart from such changes and so it is hardly surprising that within 24 hours of the temperature dropping, the incidence of cardiovascular events increases.

One study conducted between 1986 and 1996 in the United Kingdom estimated that 30–40,000 people die prematurely each year due to cold exposure[19] – several times more than the figure for heat-related deaths. Winters in Northern Europe can sometimes be harsh, but wherever you live a drop in temperature will be influential. In Negev, Israel, where summer temperatures often exceed 30° C and winter temperatures rarely drop below 10° C, the death rate from cardiovascular disease is 50 per cent higher in midwinter than in midsummer.[20]

In southern California, where the summer–winter variations in temperature are relatively small, the death rate from cardiovascular disease in December and January is 33 per cent higher than from June through to September.[21]

## Stroke

While data about stroke risk is less clear, and not all studies agree,[22] strokes are also more common in cold weather.[23] There are three main types of strokes. *Ischaemic stroke* – the most common type – accounts for almost 80 per cent of all strokes and is caused by a clot or other blockage within an artery leading to the brain. An *intracerebral haemorrhage* is caused by the sudden rupture of an artery within the brain. Blood is then released into the brain, compressing brain structures. A *subarachnoid haemorrhage* is also caused by the sudden rupture of an artery, but differs from an intracerebral haemorrhage in that the location of the rupture leads to blood filling the space surrounding the brain rather than inside of it.

Ischaemic stroke and intracerebral haemorrhage are the two types of stroke that are most often triggered by winter weather. In one study, the risk of ischaemic stroke on cold days in Russia was 32 per cent higher than on warmer days.[24] In Finland, the rate of ischaemic stroke events is around 12 per cent greater in winter than in summer, with men being most at risk. For intracerebral haemorrhage, the risk is up to 33 per cent higher, with women being most at risk.[25]

These findings are consistent with those of several other studies.[26] For example, a study from Japan found significant winter increases in the incidence of all strokes (except subarachnoid haemorrhage).[27] Another in Italy concluded that all types of strokes were more frequent during winter, with intracerebral haemorrhage rates somewhat higher in autumn.[28]

There is some evidence that it isn't just the cold but also the shorter days and relatively low humidity that link wintry weather to an increase in strokes.[29] Indeed, in a study in Israel, the average daily incidence of stroke was approximately twice as great on hot days as on relatively cold days,[30] suggesting that exposure to extreme temperatures, whether cold or hot, may, in the end, be the more crucial factor.

## How cold weather stresses the body

The biological reasons for the increase in cardiovascular events during winter are not at all clear, but several possible mechanisms have been suggested. Chronic infections (eg, infection with *Chlamydia pneumoniae* or *Helicobacter pylori*) may particularly increase the risk of stroke,[31] though evidence for this is not conclusive. Likewise, seasonal variations in air pollution, exposure to sunlight, incidence of influenza and diet may also play a role. But variation in temperature is still considered the most likely reason.[32]

Winter weather can mean an increase in levels of thyroid-stimulating hormone and cortisone, as well as in epinephrine and norepinephrine. These last three hormones are all controlled to some extent by the hypothalamus, which is central to thermoregulation, and serve to increase basal metabolism. As metabolism increases, the body begins to work harder to create heat from food.

In cold weather, the same mechanism (vasoconstriction) that helps balance core body temperature also results in a rise in blood pressure.[33] Normal seasonal variations in blood pressure means that it can be between 2–10 mm Hg higher in winter than in the summer.[34]

This change in blood pressure may influence another winter phenomenon, atrial fibrillation – or chaotic heart-beats. Atrial fibrillation is one of the most frequent causes of stroke and heart failure. The incidence of atrial fibrillation is higher in winter in cold countries such as Scotland[35] and Denmark[36] but also in more temperate regions such as Greece.[37] In the Scottish data, hospital admissions for atrial fibrillation in those aged 75 to 84 were 25 per cent higher in winter than in any other season, while among those 85 and older it was nearly 40 per cent higher.

Similarly, heart rate variability is known to be depressed in the winter.[38] Heart rate variability (HRV) is the natural rise and fall of your heart rate in response to your breathing, blood pressure, hormones and even emotions. In a healthy heart the rate should, for example, increase as you inhale and decrease as you exhale. HRV is believed to be reflective of a person's general state of health. When heart rate variability is less robust, it indicates that the body is somehow unable to respond to external and internal stress. Depressed heart rate variability is often seen in those with coronary artery disease, heart attack, congestive heart failure, hypertension and arrhythmias as well as in the presence of opportunistic infections such as herpes, colds and flu.

## Winter thickens the blood

Your blood really does 'thicken' in the winter. This is because of increases in a clotting agent known as fibrinogen.[39] Blood fibrinogen may be as much as 23 per cent higher in the coldest six months of the year compared to summer months,[40] making the blood more viscous (thicker). This is large enough to increase the risk of both heart attack and stroke in winter.[41] The elderly, in particular, can experience significant swings in fibrinogen levels during this time.[42] Fibrinogen levels also increase in response to acute and chronic inflammation, and some observers believe that minor respiratory infections occurring during the winter may be the mediator between temperature change and increased fibrinogen levels.

Levels of other components in the blood also change. The normal physiological response to cold also includes a decreased total blood volume (leading to reduced oxygen supply)[43] and a rapid change in many blood components including increased cholesterol levels (raising the risk of atherosclerosis).[44]

The winter rise in blood cholesterol levels has been so well documented in the last half-century,[45] that it is now beyond dispute. The relevance of higher cholesterol at this time of year, however, is less well investigated.

Sometimes, the difference between seasons is simply staggering. In areas where the weather changes dramatically from one season to another, such as Finland, there may be as much as a 100mg/dl seasonal variation in serum cholesterol levels.[46] ('Normal' total cholesterol levels are around 110mg/dl, so this increase could conceivably double a person's total cholesterol count.)

These winter peaks appear to be more significant for men than for women[47] and it is estimated that seasonal variations in cholesterol levels could be responsible for the 30-per-cent difference in the number of patients who are diagnosed as having high cholesterol during the winter season compared to summer.[48]

Clearly there is more to heart disease and stroke risk than cholesterol levels. Measurements of total cholesterol levels are, in particular, misleading since there is both 'bad' or LDL (low density lipoprotein) cholesterol and 'good' HDL (high density lipoprotein) cholesterol. Relative levels of each are considered more sensitive predictors of the risk of cardiovascular disease. Nevertheless, winter cholesterol increases seem significant enough to form a risk factor in some individuals.

Amazingly, seasonal variations in cholesterol are not taken into account in official management guidelines for health practitioners, nor are they mentioned in information leaflets for patients.[49] Yet, the summer/winter differences in the frequency of patients being labelled as having 'high' cholesterol carry extraordinary implications for both patient health and cost of treatment, especially given recent recommendations that adults should undergo regular cholesterol screening. Those otherwise healthy individuals diagnosed with 'high' cholesterol in the winter may well benefit from a year-long monitoring programme that builds a picture of the natural rise and fall of their cholesterol levels before going straight to drug therapy.

## The influence of diet

Although intuitively appealing, the idea that seasonal differences in diet largely explain the seasonal differences observed in blood cholesterol levels has not been well studied. Nevertheless, in one large study seasonal dietary changes accounted for only 10.5 per cent of the rise in winter cholesterol levels.[50]

There is some evidence that one gram of ascorbic acid daily can reduce or abolish the winter rise in serum cholesterol,[51] and this suggests that the seasonal variation of cholesterol levels may be related to changing intake of vitamin C-containing foods over the winter season. This would indicate that what is missing from our winter diets may be more significant than what is included.

Diet may be influential in other ways too. Recent evidence from the University of California at Irvine suggests that our dietary needs are largely determined by heredity. According to geneticist Douglas Wallace and his colleagues, mitochondrial deoxyribonucleic acid (DNA) is responsible for adaptation to different climates.[52] The mitochondria – the powerhouse of the human cell – generates energy by telling the body to eat.

After looking at differences in climate, the researchers found people in warmer climates have extremely efficient mitochondria, which use more energy for work than for producing heat. Less efficient mitochondria, such as that which people in colder climates possess, store more of the energy they receive from food to produce heat.

But this adaptive response might not be beneficial when people native to one climate relocate to another. For example, people who live in tropical or equatorial regions have a strong hereditary need for high carbohydrate diets rich in vegetables, fruits, grains and legumes. These foods tend to provide the kind of body fuel that is most compatible with an active lifestyle in warm and humid regions of the world. Their systems are simply not designed to process or utilise large quantities of animal protein and fat. Conversely, people from cold harsh northern climates are not genetically equipped to survive on light vegetarian food. They tend to burn body fuel quickly, and so need heavier foods to sustain themselves. Inuit people living in their native climate, for example, can easily digest and assimilate large quantities of protein and fat – the very types of foods that would overwhelm the digestive tracts of people from the Mediterranean basin. Either way, a move from your native climate may end up causing lifelong weight gain and a host of other related illnesses.

Such evidence throws up more questions than it answers – for instance, what climate is best for those who are 'mixed' types with no clear genetic heritage? From the perspective of the weather and health, however, it might explain why a cold snap in an otherwise moderate climate might cause such devastating health effects. Under such circumstances, the cells may simply not be able to adjust their heat-producing mechanism quickly enough.

## Physical activity

For many of us, cold weather means a drop in our physical activity levels. In a Scottish study of leisure-time activity among more than 16,000 men and women, 32 per cent reported exercising for at least 20 minutes three or more times weekly in the summer, whereas only 23 per cent exercised that frequently in the winter.[53]

When researchers at the Mayo Clinic in Rochester, Minnesota, examined seasonal variations in physical activity in 65 healthy post-menopausal women, they found that activity peaked in August and hit a low, dropping off by around 21 per cent, early in February.[54]

How this may relate to the seasonal rise in heart disease and stroke is not well studied. What evidence there is, however, suggests that cholesterol levels rise in the winter regardless of age, gender, body mass index, how active a person is or what they eat.[55]

There are, however, good reasons to remain active in winter. Sitting at home by the fire in the dead of winter is a comfortable pastime, but it may also be one reason why we are more prone to colds and flu (see below) at this time of year. There is a wealth of research that links regular moderate exercise with improved immune function[56] and reduced susceptibility to the common cold.[57]

However, exercise, like everything else, has a downside. While regular moderate exercise is helpful, heavy bursts of exercise once in a while may be harmful, suppressing immunity for several hours and creating a period of vulnerability when the risk of upper respiratory tract infections is increased.[58] Interestingly, the soldiers most severely affected by the US swine flu outbreak in the winter of 1976 were just beginning basic combat training, a time of exceptional exertion.[59]

* * *

## Seasonal Affective Disorder (SAD)

While the concept of a 'winter depression' was met with scepticism when it was first mooted in the early 1980s, it is now a widely accepted fact. Known formally as Seasonal Affective Disorder (SAD), it is a condition characterised by autumn and winter depressions, alternating with non-depressed periods in spring and summer. During their winter depression, SAD sufferers experience symptoms of fatigue, carbohydrate craving[60], overeating[61], weight gain and oversleeping on top of the emotional experience of depression.

SAD sufferers have unusually high levels of melatonin – a hormone that regulates our sleep–wake cycle. Production of melatonin is highest at night or when it is dark and in SAD sufferers the shorter days and longer nights of winter can lead to an increased urge to sleep and/or a reduced desire to interact with others.[62] Melatonin production generally decreases with age, but the elderly do experience pronounced seasonal shifts and SAD may worsen with age.[63] SAD is also significantly more prevalent in northern regions where daylight hours are generally shorter.[64] In fact, the nearer you live to the North Pole the more common it is.

Moving south can help. Bright artificial light has also been shown to reverse the symptoms of SAD, including carbohydrate craving[65] and early morning exposure to light appears to be more effective than exposure late in the day.[66]

SAD undoubtedly makes coping in winter all the more difficult. But new evidence suggests that SAD may simply be an extreme manifestation of what all of us experience in the winter. Studies show that increased weight, appetite and

## Schizophrenia and mood disorders

Is schizophrenia caused by mothers being deprived of sunshine during pregnancy? That is a question that has been debated for many decades. Babies born between February and April are 10 per cent more likely to develop schizophrenia than those born at other times of the year.

There is, of course, no one cause for schizophrenia; genetics and environment also play a part. Nevertheless, season of birth accounts for more cases of schizophrenia than many other known risk factors – including genetics.

Only a city birth is a higher risk factor than being born in the spring. This is because of the relative indoor lifestyle of city dwellers. One of the things UV does is convert a cholesterol-like molecule in the skin to vitamin D. Vitamin D is low in winter, low in cities and low in dark-skinned migrants to northern climates – all high risk factors for schizophrenic births.

The Southern Hemisphere gets around 15 per cent more UV light than the Northern Hemisphere, which the scientists say would explain the generally higher rates of schizophrenia in Europe and America.

It is not known what role vitamin D might play in the development of the brain, though experiments suggest it may be necessary for building the brain and developing tissue.

It has also been proposed that mothers who had had an infection, such as flu, during pregnancy at the time the babies' brains were being formed were more likely to have children who go on to develop schizophrenia. However, Australian evidence suggests that it could be the lack of UV light that is the key factor. Dr John McGrath of the Queensland Centre for Schizophrenic Research found that in Queensland there is a peak in the birth of schizophrenic babies every three to four years.[69]

This does not coincide with the big flu outbreaks, but does happen with the same frequency that the El Niño weather system gives that area extremely gloomy weather and therefore a lack of UV light.

carbohydrate craving in autumn and winter are common among the general population.[67] So are changes in both mood and behaviour.[68]

It may even be that the symptoms of SAD are a form of primitive protection. In our caveman past, semi-hibernating when there wasn't much food around could have been a useful way of getting through winter. Even now, having a natural low, especially in winter, may be important for recharging our batteries and enabling the natural 'high' that is a common experience in spring and summer.

## Respiratory illness

Winter is commonly referred to as the cold and flu season. Beginning in late August or early September, the incidence of colds begins to increase slowly for a few weeks and remains high until March or April, when it declines. But although this 'cold season' tends to correspond with winter weather, you can't actually catch a cold from being exposed to cold.

Instead, the sneezing, scratchy throat and runny nose that everyone knows as the first signs of a cold are influenced by the weather in other, less direct ways.

More than 200 different viruses are known to cause the symptoms of the common cold. Rhinoviruses cause an estimated 30 to 35 per cent of all adult colds though they seldom produce serious illnesses. These viruses are most active in early autumn, spring and summer. More than 110 distinct rhinovirus types have been identified and these proliferate best at temperatures of 33 °C/91 °F – the temperature of the human nasal mucosa.

Coronaviruses (the same virus responsible for the recent SARS 'epidemic') are believed to cause a large percentage of all adult colds. Coronaviruses induce colds primarily in the winter and early spring. Of the more than 30 strains isolated so far, three or four infect humans. Our understanding of coronaviruses and how they relate to the incidence of cold and flu, however, is incomplete because, unlike rhinoviruses, they are difficult to grow in the laboratory.

Approximately 10 to 15 per cent of adult colds are caused by viruses which most of us have never heard of and which are more typically responsible for other, more severe illnesses: adenoviruses, coxsackieviruses, echoviruses, orthomyxoviruses (including influenza A and B viruses), paramyxoviruses (including several parainfluenza viruses), respiratory syncytial virus and enteroviruses. Other viruses that we presume account for as many as 30 to 50 per cent of adult colds are still unidentified.

## Catching a cold

Weather and climate can make or break a flu season. Shortly after the first frost, the first few cases of flu will develop. There is good evidence that flu outbreaks are more serious in cold and very cold weather conditions. In the winter of 1996–97 in the UK (a very cold winter), statisticians estimate that 49,000 more deaths occurred than expected. These are what scientists call 'excess deaths'; in other words, they are not connected with heat, cold or accident caused directly by the weather (a cardiac arrest, while shovelling snow, for example, is a weather-related accident and not considered an excess death). Many of these excess deaths were due to influenza, which peaked in December and January and coincided with the coldest temperatures.[70]

Catching a cold or flu is not a random event. We may live, work and take our leisure in poorly ventilated, sometimes overcrowded environments that encourage the concentration of virulent viruses.

Washing your hands, covering your mouth when you sneeze or cough and not soldiering on to work or school when you are ill, are all important ways of avoiding spreading or catching colds. But if exposure were the only factor, each of us would get sick every time we were exposed.

Many people in a room can be exposed to the same virus but only some will become infected. Susceptibility, not exposure, is the key and this is likely to be influenced by a number of things.

People who are ill with heart complaints, asthma, chronic kidney disease or diabetes or who are taking medications (for instance, steroids) are, of course, more susceptible to colds and flu. People who smoke are also more susceptible to upper respiratory infections.[71]

But other things also influence our susceptibility of viral infections. Winter weather is inherently stressful and the ability of stress to alter immune function and to precipitate and aggravate infectious diseases has long been recognised.[72] Today, evidence is strong that the single biggest risk factor that puts otherwise healthy people at risk of catching a cold may be stress.

Stress – physical, emotional, psychological or spiritual – can lower your natural immunity[73] and effectively doubles your chance of getting a cold,[74] as does lack of social support and human interaction.[75] In the same vein, depression – also common in the winter months – is strongly related to catching a cold.[76]

## Respiratory distress

While hay fever reactions drop right off in winter, colder weather can raise levels of other types of spores such as moulds. Many studies from around the world have shown that dampness in the home can increase respiratory and other allergic reactions.[77] This is true for both adults[78] and children.[79] Fog, dew and frost, especially on days when the temperature is cool and there is relative calm, can increase the amount of solid particles – man-made pollution such as soot and industrial emissions but also fungal and mould spores – in the air and this has been mooted as a cause of winter allergies.[80]

Exercise-induced asthma (EIA) is another good example of a winter-related respiratory problem. It is especially common in children and young adults, and comes on most often during intense exercise in cold dry air. EIA is not necessarily linked to an allergic reaction and you don't have to be an asthmatic to suffer from it, though many asthmatics do have this problem.[81] Neither do you need to be exercising to experience this kind of respiratory reaction to cold air – any unaccustomed exertion can bring on a similar reaction.

The main trigger for EIA appears to be inhalation of very cold, dry air that cools the mucous membranes of the upper respiratory tract and may eventually cause irritation, micro-inflammatory reactions and produce bronchospasm.

When we are engaged in aerobic exercise, there is a large and sudden increase in the volume of air that the body must humidify (i.e. moisten) and warm. The airways in people prone to EIA tend towards hypersensitivity and may have thermally sensitive neuroreceptors.[82] This means that their body reacts to the cooling and drying of the respiratory tract during aerobic exercise with bronchoconstriction – a contraction of the smooth muscle around the bronchioles. This leads to the symptoms of EIA, such as tightness in the chest, coughing, wheezing and shortness of breath.

It is not entirely clear why cold, dry air should have this effect on some individuals; however, one theory is that once

### A cancer climate?

Digestive tract malignancies seem to occur more often in colder climates.[85] There is also a seasonal fluctuation in the detection of human papilloma virus and cervical cancer (in fact the two are related);[86] cases detected in February are almost double those of any other time of the year.[87] The peak incidence of newly diagnosed uterine cervical cancer cases occurs in February. No clear causative factor has been identified.

Prostate cancer, breast cancer and colon cancer have all been implicated in a cancer belt that stretches across America's more cloudy climates.[88] Rates for these cancers are 2–3 times higher than incidences of cancer found in sunnier locales.

Similarly, in a study from New Zealand, women with previous breast cancer had an abnormal melatonin rhythm, with an unusual drop in wintertime malatonin detected in those who were diagnosed with breast cancer in that season.[89]

Again the problem may be lack of vitamin D. If you live in perpetually cloudy or winter-like conditions, low vitamin D levels may present a genuine threat to health. One of vitamin D's functions is to regulate the development and replication of cells[90] and laboratory experiments have shown that vitamin D blocks the proliferation of breast, colon and prostate cancer cells (see Chapter 6 for more).

in the respiratory passageways, this low-moisture air triggers the release of mast cells (inflammatory mediators) and this in turn leads to bronchoconstriction.[83]

Although EIA is not generally considered an allergic response, there are some who believe that allergens, in the form of air pollution and particulate matter, may have a role to play in triggering bronchoconstriction.[84]

In people with an already weakened cardiovascular system, bronchospasm – difficulty breathing and a possible coughing fit – can place extra burden on the heart leading, in some cases, to heart attack or stroke.

## Winter skin

The first change you may notice in your skin in winter is its colour. Winter pallor is the result of blood being drawn away from the skin in order to maintain your internal heat balance.

Winter weather also brings with it a whole range of other skin disorders from chilblains to urticaria.[91] Women, in particular, are prone to chilblains, painful swellings, patches or blisters on the feet, hands, face and ears caused by exposure to the cold. Once these develop, they will return again and again with repeated exposure to cold.

Human skin is composed of several layers of cells that are partially protected from dehydration by an insulating blanket of oil. Once those oils are stripped away, either from too much washing, the modern obsession with 'exfoliating', or by cold, windy weather, the skin begins to dry out. Skin, of course, is a barrier. It is designed to keep harmful things out as well as keeping healthful things in. When skin begins to crack and get sore, it allows bacteria and other microbes to penetrate this protective barrier.

While dry skin can be a problem at any time of the year, cold weather can dry out your skin in two ways: directly, because cold air generally has low humidity, and indirectly, from central-heating systems. Dryness is a very common skin problem and is often worse during the winter when environmental humidity is low (thus the name, 'winter itch'). To prevent this, you may need to change the way you treat your skin in the winter, not washing with water that is too hot or too cold, using good quality moisturiser applied onto damp skin and a humidifier to restore moisture to home or office air.

Finally, although we think of cold sores (caused by the herpes virus) as a winter phenomenon – and indeed they may break out if you catch a cold in the winter – these painful blisters are actually most common in the summer when exposure to sunlight can act as a powerful trigger.

## Extreme weather – snow and ice

Winter storms bring can bring snow, sleet, freezing rain, fog and ice. While snow, in particular, can make the world look fresh and new, and bring out a childlike sense of awe and fun in most of us, it can also be a tremendous health hazard.

Snow blows into our lives via low-pressure systems that sweep across the continents in the mid to high latitudes. In some areas, snow is an occasional treat, in others it is an integral part of life and everything from transport to housing and food depends on when and where the snow falls.

With snow and ice, the risk of accidents increases. In the cold weather, our manual dexterity is reduced. People who work with their hands, either on an assembly line or driving a car, may find their manual dexterity is greatly reduced in cold weather. Winter clothing is also a hazard. Gloves make simple jobs more awkward. Winter coats can be bulky and restrictive, and hoods and hats reduce our field of vision. These and other winter hazards can cause injury and death.

The way we heat our homes in the depth of winter is also important. Any process of combustion can produce carbon monoxide (CO). Commonly, it is emitted from combustion engines (automobiles, gas-powered lawnmowers, etc) but poorly ventilated kerosene or space heaters, furnaces, woodstoves, gas stoves and fireplaces also produce significant amounts of CO. Thus, in winter, deaths from carbon monoxide poisoning rise.

Once inhaled, CO inhibits the distribution of oxygen in the blood to the rest of the body. Depending on the amount inhaled, the symptoms include fatigue, headache, weakness, confusion, disorientation, nausea and dizziness. These symptoms are sometimes confused with the flu or food poisoning. Children, the elderly and people with heart and respiratory illnesses are particularly at high risk for adverse health effects of carbon monoxide. Long-term exposure to carbon monoxide is associated with heart disease; extremely high levels can cause death.

## Hypothermia

During the winter, the body has to work harder to maintain its constant temperature (another reason why winter brings greater tiredness and lethargy) and hypothermia – a dramatic loss of body heat – is more common in winter than most of us realise. Very simply, you are suffering from hypothermia when your body loses heat more rapidly than it can replace it. While it is not solely linked to snow or ice, a prolonged phase of extreme cold weather inevitably brings with it hypothermia fatalities.

Hypothermia occurs when the core body temperature falls below 35° C/95° F.[92] Certain individuals are more susceptible to hypothermia than others, including the elderly, newborns, alcoholics, the unconscious and people on medications.[93] In addition, malnourishment and inadequate housing also increase the incidence of hypothermia.

The elderly are particularly at risk because vasoconstriction and shivering, the body's primary adaptive measures to cold, are less effective in many older people.[94] In addition, many elderly people do not discriminate changes in temperature well and are thus less able to adjust to them.[95]

Sex and race also appear to influence our susceptibility to hypothermia. Non-white elderly men are among those most at risk, while white women appear to be at the lowest risk.[96] Women in general appear to be better able to maintain a higher body core temperature during periods of cold stress,[97] and some studies suggest that this is because women's bodies naturally contain a larger amount of fat – though others disagree.[98] Although women are less susceptible to hypothermia, they appear to be more susceptible to peripheral cold injuries such as frostbite.[99]

As with extremes in hot weather, there is a 'lag time' of two to three days between the onset of cold weather and a sharp increase in the number of hypothermia-related deaths,[100] suggesting that the extreme change in temperature triggers a spiral where the body grows gradually weaker and less able to maintain a balance in the face of external weather pressures.

### Alcohol and winter don't mix

Many of us believe that a little 'nip' of something on a cold winter's day will help take the chill away. Initially, this appears to be the case. A drink of alcohol brings colour to our cheeks and makes us feel a little warmer, maybe even makes us perspire. However, these temporary and superficial signs of warmth hide the chilling facts about drinking alcohol in cold weather.

Consumption of alcohol is completely counterproductive to the body's own cold-survival mechanisms. Alcohol causes a sharp decrease in core body temperature by causing the blood vessels to dilate. This sends blood to the superficial layers of skin and tissue (thus the glow in our cheeks) where it cools rapidly. As this blood cycles back into the deeper tissues of the body, it causes the core temperature to drop rapidly, increasing the risk of hypothermia.

## Frostbite

Allied to hypothermia is frostbite. The world's highest mountain peaks have been climbed without cold injury and the military engages in manoeuvres in cold environments with a minimum of cold injuries. Yet, for average citizens, frostbite is still a reality. Frostbite injury is most common in adults aged 30 to 49 years, and studies reveal that the feet and hands account for 90 per cent of all reported frostbite. Other vulnerable sites include the ears, nose, cheeks and penis.

Recently, concern about frostbite injury has become more widespread. Our increased participation in outdoor activities and sports in cold environments has contributed to a rise in the rates of frostbite injury.[101] So has the rise in homelessness in the West.[102]

Studies from around the world show that there are several risk factors for frostbite injury.[103] A 12-year study tracking injuries to 650,000 people living in Saskatchewan in Canada found that alcohol consumption, motor vehicle problems and psychiatric illness accounted for 46, 19 and 17 per cent respectively of frostbite injuries.[104]

Other studies from Norway, Finland, and the United States have linked frostbite to homelessness, improper clothing, a history of previous cold injury, fatigue, wound infection, atherosclerosis, diabetes and smoking.[105] Smoking hastens cold damage in a different way to alcohol (see box page 163) by increasing vasoconstriction, thus making cold injury at skin level likely to occur quickly.

In contrast, acclimatisation, general good health, exercise, normal blood pressure and adequate protective clothing can help prevent frostbite even in extreme weather.[106]

A related problem, Raynaud's disease, affects the blood flow to the fingers, toes, ears, lips and nose. While not limited to seasons or areas with cold climates, Raynaud's can be quite serious. In Raynaud's sufferers, the sensitivity to cold is almost too great and vasoconstriction too intense, sometimes cutting off blood supply completely to the affected area leaving it icy cold and waxy looking. If this goes on for too long, tissue death (gangrene) can occur, thus it is particularly important that Raynaud's sufferers cover up well in cold weather.

# Chapter Eight

## Artificial Environments

The popular conception of human progress holds that each new stage of history has been an improvement, bringing technological breakthroughs that have raised the quality of human life. However, one of the biggest shifts in human evolution – from living a life exposed to the elements to living largely sheltered from them – may not have improved our life and health as much as we believe. Today, most human beings spend 80–90 per cent of their time indoors. This transition from being outdoor creatures to indoor creatures has been made in a relatively short space of time and has brought both benefits and disadvantages.

Life indoors can be more comfortable, especially when the weather outside is either very hot or very cold. Indoors we are largely in control of our own environments, able to make use of advances in technology to adjust temperature, humidity, lighting and the flow of air to suit our individual preferences. Since the oil crisis of the 1970s, we have worked hard to make our buildings energy efficient, for instance, with double glazing and sealed windows – thus uncomfortable drafts are largely a thing of the past in most modern buildings. We also use a range of fabrics and insulation materials indoors to help keep warmth in and to ensure our comfort.

Our indoor environments, in fact, represent a valiant attempt to do nature one better – to create what has been called a 'uniclimate' – one that is predictable and where the capricious highs and lows of the natural environment are completely eliminated.

Under the right conditions, a stable indoor environment can certainly increase well being and productivity. But life indoors also means less access to those things we need the most to remain healthy such as fresh air and sunshine. It is a more sedentary lifestyle and, ironically, more likely to bring us in contact with toxic substances that ruin our health.

Buildings are complex environments. In addition to trapping heat and air, they can also trap as well as generate a range of indoor pollutants, for instance:

- ***Volatile organic compounds*** – neurotoxic gases including benzene, styrene, tetrachloroethylene and paradichlorobenzene – emitted from building materials, furniture, soft furnishings, wallpaper and paint, air fresheners, perfumes, plastic storage and plastic casings (e.g. on computers), photocopiers and franking machines, among others.
- ***Combustion by-products*** – the result of incineration or burning, thus cigarette smoke, emissions from heating units, even human metabolism and breathing (combustive processes that generate heat and gases such as carbon dioxide).
- ***Respirable dust and particulates*** – including heavy metals, asbestos and fibreglass.
- ***Bioaerosols*** – living and non-living biological contaminants such as microbes (bacteria, mould, fungi), animal dander and dustmite allergens.
- ***Radiation*** – energy emitted from electrical equipment, as well as naturally occurring radon gas.

The price we pay for the indoor uniclimate is high; often unsafe levels of each of these pollutants. The effects on human health are only just beginning to be revealed.

## Artificial air

Air is life. Every time we inhale we absorb oxygen, vital to the optimum function of every cell, tissue and organ in the body.

But along with life-giving oxygen we are also absorbing a range of hazardous gases and minute particles of toxic dust. In fact, whatever is in our air is also in our bodies. So the quality of the air we breathe is vitally important.

Fears about outdoor pollution have increased in recent years and worries about rising rates of asthma and possible links with skin cancer make many people feel like hiding away indoors. Indeed, many parents keep their asthmatic children indoors on particularly polluted days in order to avoid pollution-triggered asthma attacks. Unfortunately, as evidence from the US Environmental Protection Agency (EPA) and other governmental agencies around the world has shown, the air indoors – in our homes, offices and even schools – is far from safe.

In 1985, the EPA published the findings of one of the most famous and devastating studies into indoor air quality. The Total Exposure Assessment Methodology (TEAM) study[1] changed most of our cherished assumptions about the safety of indoor air by showing that our greatest personal exposure to pollutants, in particular carcinogenic and neurotoxic volatile organic compounds (VOCs), is from air inside the home and not from outside air as had previously been thought.

The EPA scientists looked for the presence of twenty VOCs in samples of indoor air, outdoor air, personal air (that is the air very close to each individual, measured by monitors attached to clothing) and respired air (in the lungs) in a total of 780 people. Personal air samples contained the highest levels of VOCs – much higher than could have been predicted from any previous data. More worrying, levels of VOCs in the personal air space were much higher at night – when the body should be resting and repairing – and once again much higher than those measured outdoors during the same time frame.

As part of our commitment to creating the perfect uniclimate, many of us 'bring the outdoors in' by using 'natural' scented air fresheners. The TEAM data found that the neurotoxic solvent dichlorobenzene, commonly found in air fresheners, was detected in high concentrations in indoor air. Ten years after the TEAM study was published, another US study found this pollutant in urine samples of 2 per cent of all children in one US state and in 98 per cent of 1,000 selected adults from across the country.[2] Another found that dichlorobenzene, and three other toxic solvents – xylene, ethylphenol and styrene – also given off by air fresheners but in addition present in tobacco smoke and nearly every modern building material, were present in 100 per cent of tissue samples tested across the country.[3]

In the UK, the picture is much the same. In 1996, the Building Research Establishment, in conjunction with the Medical Research Council's Institute for Environment and Health, published data based on monitoring 174 homes in Avon in the West of England. They found that levels of formaldehyde gas were ten times higher indoors than out. In addition, 12 homes exceeded World Health Organization recommendations for indoor air quality. According to the researchers, the levels of indoor pollution were due to the cleaning agents used as well as gases generated by modern appliances, for instance carbon monoxide, benzene and VOCs.[4]

Numerous studies conducted by the EPA since the TEAM data was released have shown measurable, and in some cases disturbing, levels of more than 100 known VOCs in modern offices and homes. In addition, reports from other countries, such as Australia[5] and Canada, also confirm that levels of VOCs and other harmful pollutants tend to be higher indoors than out. A recent European review – compiled a sorry 17 years after the original TEAM study – confirms that indoor air can be at least twice as polluted as outdoor air, and that dangerous chemicals are regularly emitted from carpets, furnishings, paint and computers. The report from the Joint Research Committee confirms that up to 20 per cent of Europeans suffer from asthma due to substances inhaled indoors, and that tobacco smoke, asbestos, radon and benzene released inside buildings are prime suspects in the increase in cancer cases amongst the European population.[6]

## Sick Building Syndrome

Inhaling air with high levels of these indoor pollutants is largely responsible for a disorder known as sick building syndrome (SBS). People suffering from SBS often experience debilitating symptoms such as headache, dizziness, disorientation, difficulty concentrating, fatigue and eye, nose and throat irritation.[7] Buildings that are sealed up tight against the natural environment are much more likely to produce these symptoms in their inhabitants. Indeed sick building syndrome was originally called 'tight building syndrome'.[8]

Sick building syndrome is a complex problem and for convenience its associated symptoms are generally split into one of two broad categories: sick building syndrome or humidifier fever. The former is usually associated with chemical off-gassing, while humidifier fever is usually associated with overgrowth of bacteria in air-conditioning and humidifier units. These are further divided into type 1 and type 2 symptoms. Type 1 symptoms are similar to symptoms of cold and flu; type 2 symptoms tend to be closer to allergic reactions. Each type has its own unique symptoms:

- ***Sick building syndrome (type 1)*** – lethargy and tiredness, headache, dry, blocked nose, sore, dry eyes, sore throat and dry skin and/or skin rashes.
- ***Sick building syndrome (type 2)*** – watering/itchy eyes and runny nose, symptoms similar to hay fever.
- ***Humidifier fever (type 1)*** – flu-like symptoms, generalised malaise, aches and pains, cough, lethargy and headache.
- ***Humidifier fever (type 2)*** – chest tightness, difficulty in breathing, fever, headache, wheezing and occupational asthma.

These short-term effects can be dramatic and debilitating enough. But some observers believe that exposure to these toxic gases over the longer term may be a risk for more serious problems.

For instance, according to a recent EPA report, *Healthy Buildings, Healthy People: A Vision for the 21st Century*, the long-term risks associated with poor indoor air quality include asthma, cancer, reproductive and developmental problems, among others.[9]

The report confirms that dust mites and other allergens, micro-organisms, second-hand smoke and some chemicals found indoors are triggers for asthma. It also notes that a number of indoor contaminants, such as asbestos, radon, tobacco smoke and benzene, are known human carcinogens. Other indoor contaminants, such as certain chlorinated solvents, polycyclic aromatic hydrocarbons, aldehydes and pesticides, are considered likely to cause cancer in humans.

Many chemicals commonly found indoors (e.g. tobacco smoke, some pesticides, lead and other heavy metals, alcohols and plastic additives) are hormone disrupters capable of causing reproductive problems in humans including miscarriage and fetal abnormalities.

Indoor environments can cause or amplify many other health problems as well.[10] Several studies show that exposure to passive smoke, even occasionally, raises the risk of developing cardiovascular diseases including hardening of the arteries and heart attack.[11] The increase is dramatic – between 25 and 35 per cent greater than for those not exposed to cigarette smoke.[12]

Carbon monoxide (CO) poisoning, associated with the improper use and maintenance of fuel-burning appliances, can cause illness and in some cases can be fatal.[13] The organism that produces Legionnaires' disease, a potentially deadly form of pneumonia, lives in cooling systems, whirlpool baths, humidifiers, food market vegetable misters and other indoor sources, including residential tap water.[14] Effects associated with toxins from indoor fungi and bacteria range from exacerbation of asthma to rhinitis, conjunctivitis, recurrent fever, lethargy and malaise, difficulty breathing, chest tightness and cough.[15]

## Artificial temperatures

Central heating and air conditioning have become so ubiquitous in modern buildings that we hardly give them a second thought. These are, in fact, the main tools we use to create the indoor uniclimate.

As already illustrated in this book, both heating and air conditioning may greatly interfere with the way we adapt to outdoor temperatures. In both summer and winter, the transition from the uniclimate indoors to the real one outdoors may mean an abrupt change in temperature of 10° C/18° F or more.

Hot or cool air forced through the duct work of most central heating and air-conditioning systems also sets up friction that results in an overabundance of positive ions. Synthetic materials used in modern offices add to this problem and the result can be Serotonin Irritation Syndrome (see Chapter 4) that can upset both the mental and physical equilibrium of everyone, not only those who are ion sensitive. Positive ion overload is thought to be a contributing factor to the symptoms of sick building syndrome.

Some may view such symptoms as a small price to pay if it means keeping us alive in weather extremes. But it is unclear whether modern contrivances such as central heating and air conditioning actually do save lives. Of the two, air conditioning and other types of ventilation systems do appear to make a difference on very hot days,[16] although no one can say for sure how much these reduce heatwave-related deaths. While the proportion of households with central heating in England and Wales has increased substantially since the 1960s, this has not been accompanied by an acceleration in the already declining trend in excess winter mortality since the 1940s.[17] In one 1980s study, providing unrestricted central heating to elderly residents of housing association homes had no impact on winter mortality.[18]

### Heaven at the mall

Research from around the world has shown that every type of weather affects our shopping habits, inducing us to either spend or save. In New Zealand, as rainfall increases, the number of shoppers decreases. In Sydney, Australia, where temperatures are higher, shoppers flock to malls during hot and humid weather to take advantage of the air conditioning. Edmonton, Canada sees people relishing the indoor warmth of the world's largest mall during the winter months of snow and sub-zero temperatures.

This understanding is now being used by architects and retailers to entice shoppers into the shopping mall in all weathers, for instance by providing air conditioning in summer and valet parking for open car parks commonly subject to wet weather.[19]

What artificial temperature control most certainly does, however, is make the problem of indoor air pollution worse. Overall, higher indoor temperatures tend to increase SBS symptoms, while lower temperatures reduce them. Most people know this intuitively. Studies show that office workers perceive indoor air to be better when temperature and/or humidity are towards the low end rather than the high end of the comfort zone.[20]

Likewise, heating combined with heat-conserving measures such as sealed windows causes an increase in eye, nose, skin and respiratory problems.[21]

Asthma-related illness is one of the leading causes of school absenteeism, accounting for over 10 million missed school days per year.[22] A recent European study of 800 students from eight different schools, suggests a very direct link between heating, lack of ventilation and health problems and students' ability to concentrate.[23]

In classrooms where carbon dioxide levels were high (caused by crowding, poor ventilation and people breathing – we exhale $CO_2$), student scores on the concentration tests were low; and reported health symptoms were high.

Other evidence shows that turning up the heat even modestly affects children's abilities to perform mental tasks requiring concentration, such as addition, multiplication and sentence comprehension.[24]

Research involving adults has shown that exposure to pollution sources – in this case a 20-year-old carpet that was introduced into the room without the participant's knowledge – in a heated environment, significantly affected concentration and accuracy at work.[25]

Air-conditioning systems, which can either heat or cool fresh and recirculated air, do not contribute to reducing indoor air pollution, but they do alter indoor air in other ways. Evidence consistently shows that complaints about indoor air quality and symptoms of sick building syndrome are significantly more prevalent in air-conditioned rather than naturally ventilated offices.[26]

Ventilation systems usually re-circulate around 80–90 per cent of indoor air, bringing in 10–20 per cent of fresh air from outside. This means that the majority of gases already present in the indoor environment remain in the atmosphere. This is a problem that can't be cured by simply circulating more air at a faster rate. It has been calculated that to clear the air in an average indoor environment would require the equivalent of a tornado-like gale to be circulated throughout the area.[27]

In addition, moisture problems due to rain, water or snow are common in outdoor air intakes of many ventilation systems.[28] Via these air intakes, bird droppings with infectious fungi, *Histoplasma* and *Cryptococcus*, may cause severe health problems for the building occupants.

Air filters, porous insulation materials and materials used in modern buildings are susceptible to development of fungal growth if sufficient moisture is available. The risk is highest in humid climates where air conditioning is common. Under such circumstances contaminated insulation material may contain millions of spores per square centimetre of material.[29]

## Artificial light

Mankind evolved under the solar spectrum. Yet today most of us work and live in the feeble glow of incandescent light bulbs or under harsh fluorescent lights.

Incandescent light bulbs produce relatively little light for the energy they consume. In contrast, they produce heat in the infrared spectrum – and the light we see is really the thermal radiation emitted by the bulb – but no UV radiation. Fluorescent lighting, which uses a non-thermal radiation process to produce light, is much brighter, producing light energy in the visible spectrum, and does produce some UV light, but not in the balanced state we would normally find in nature.

Tungsten halogen bulbs are highly efficient light sources and are more and more common in home and office lighting. The filament operates at high temperature, causing it to generate visible light and significant amounts of UV radiation. For this reason, most halogen appliances are fitted with UV filters. Those that are not may emit unsafe amounts of UV radiation. The use of unfiltered desktop lamps should certainly be discouraged if they are used for more than 2 hours per day and are sited within half a metre of the user.

What this means is that life indoors effectively deprives us of the full spectrum of health-enhancing light. As John Ott, a scientist who has spent his life investigating the effects of artificial light on humans, has observed, artificial distortions of that spectrum can lead to malillumination, a condition he believes is analogous to malnutrition, and that may produce undesirable biological effects in humans.[30]

The answer to our malillumination, according to Ott, is to use full spectrum lighting indoors. It's a good theory, although most of the scientific research has been done on school children. A 1973 study in Florida, performed on young children, found that hyperactive children calm down and academic levels go up when full-spectrum lights are installed.[31] In a similar study in Canada in the 1980s, researchers noted marked behavioural improvement among children under full-spectrum lighting as well as decreased stress levels, identified by drops in systolic blood pressure averaging 20 points per child. When the full-spectrum lighting was changed back to the original cool-white fluorescent tubes, the children's stress levels shot back up and they became disorderly again.[32]

German research has revealed that adults working under cool-white fluorescent bulbs have high levels of stress hormones – specifically, adrenocorticotrophic hormone (ACTH) and cortisol. For this reason, cool-white fluorescent bulbs are banned by law in German hospitals and medical facilities.[33]

Based on animal studies, it has been suggested that children schooled under full-spectrum lights might even develop one-third fewer cavities compared to children schooled under the standard fluorescent lights.[34] This effect has been put down to the way that full-spectrum lighting increases both vitamin D levels and calcium absorption.[35] Unfortunately, this hypothesis has never been tested in human children.

Such data is encouraging, yet there are several reasons why the full-spectrum bulb has not been adopted as standard. Cost is one – full-spectrum bulbs are hard to find and much more expensive than other bulbs. But the main problem, ironically, is fear of UV radiation. UV is, of course, one of the most biologically active and important portions of the electromagnetic spectrum and we avoid it completely at our peril.

As with all things, it is a matter of education and balance. At the moment, we still tend towards lumping all forms of UV together even though each type – UV-A, UV-B and UV-C – is very different. UV-A is responsible for tanning; UV-B is crucial to the body's synthesis of Vitamin D; and UV-C is a potent germicide. Of these, UV-C presents the biggest threat to our well being since the thinning of the ozone layer means that we are receiving a much larger dose of this type of radiation than we ever have.

While official agencies continue to feed the public misinformation with regard to the relative risks of sunlight, other studies into the dangers of fluorescent lighting remain unpublicised. One published in the British medical journal, *The Lancet*, reported that a group of Australians who worked all day under fluorescent lights had higher incidents of skin cancer than people who frequently sunbathed or worked outside.[36] Officialdom baulked at the suggestion and, while little quality research has been funded to replicate the findings, some scientists believe the risk is real.[37] However, official reports on tungsten halogen light suggests that the UV radiation exposure caused by some desktop tungsten halogen lamps can be comparable with levels of solar UV radiation in terms of its ability to produce both erythemal (skin rash) and potential carcinogenic effects.[38]

Laboratory studies suggest that we need all three forms of UV radiation in balance to remain healthy. For instance, when scientists at the University of Wyoming exposed paramecia (one-celled, freshwater organisms) to the bactericidal form of ultraviolet light (UV-C), it caused the expected DNA damage and shortened the paramecia's lifespan. What was unexpected was the way that subsequent exposure to UV-A healed the damage and even extended the cells' expected lifespan.[39]

Human studies suggest that exposure to the full spectrum of UV light lowers blood pressure, improves electrocardiogram readings, reduces cholesterol, aids weight loss and the healing of psoriasis and increases the levels of male and female sex hormones.[40]

It seems incredible that simply changing the way we light our homes, offices and schools could make such a substantial difference to our health – reducing stress and potentially healing allergic skin disorders. Indeed, it has never been tested in any large-scale trial. Yet the theory is credible and such is our need for natural light that those suffering from these disorders, which may be part of a sick building syndrome, might consider making the change from normal incandescent or fluorescent bulbs to a more natural spectrum of light indoors.

## Night shifts

People working nights and rotating shifts may also be suffering from light-related health problems. Research has shown that the human body needs at least three days, and sometimes as many as seven, to adapt fully to a six-hour time shift. Shift workers, especially those who have rotating shifts, may never get a chance to reset their body clocks sufficiently. On the other hand, those permanently working night shifts, may never have a chance to adjust. Under such circumstances, the suprachiasmatic nucleus (SCN) becomes out of synch (see Chapter 2) and the individual lives in a state of perpetual jet lag.

These problems can be alleviated to some extent by educating shift workers in how to shift their sleep–wake cycle so that it is synchronised with their work schedule. Making sure that work environments are lit brightly enough to suppress melatonin production and keep workers alert is also important.[41]

These simple things, however, are very often not done and the oversight has implications not only for the health of these individuals, but also for the well being of communities as a whole. Millions of people throughout the world work in shifts or through the night and many hold jobs of critical importance such as emergency services personnel, police, air traffic controllers, doctors, nurses and power plant operators. That these workers often cannot perform at their full mental and physical capacities has been thoroughly documented.[42]

## Buildings change the weather

The climate inside our offices, homes and schools is undoubtedly influential to health. But buildings affect the weather outside as well. Due to a combination of factors, cities are generally 3–6° C/5.4–10.8° F warmer than surrounding rural areas. Urban areas and cities tend to be constructed of impermeable substances such as concrete and asphalt that prevent evaporative cooling and conduct heat more efficiently than land and vegetation.

Other factors that increase city heat include the canyon structures of high-rise buildings as well as waste heat from buildings, cars and industrial processes. The effect of this urban heat means that cities tend to be somewhat warmer in winter, but much hotter in the summer. This increases the risk of heatstroke, heat syncope and heat cramps, as well as exacerbating many pre-existing health conditions. It also increases the risk of heat-related deaths.[43] Most of the summer fatalities related to heatwaves happen in cities and are attributed to the way that city structures trap heat at night. At night, the temperature of the human body takes a natural dip, which allows us to sleep but also is part of a complex process that initiates our repair mechanisms and maintains immunity. High night-time temperatures mean the body never really rests or repairs properly.

Individuals at the highest risk of heat-related illness include the elderly,[44] those on certain medications[45] and those with pre-existing illnesses.[46] All these effects are much more pronounced in city dwellers.[47]

The weather effects of these urban 'heat islands' go well beyond hotter days and nights. One study of 17 urban heat islands in New England found the expected rise in temperatures[48] on average 1–2° C/1.8–3.6° F higher than in surrounding rural areas. But city structures also altered other atmospheric parameters; precipitation was 5–15 per cent higher in the city; there were 16 per cent more thunderstorms and 5–10 per cent more clouds. In contrast, wind speed, solar radiation and relative humidity were significantly reduced.

Air pollution also blocks the UV rays responsible for triggering the production of vitamin D in the body. The recent re-emergence of rickets among children and higher rates of osteoporosis in adults has been blamed on a combination of our largely indoor lifestyles and outdoor pollution.

In addition, urban environments create as well as trap air pollutants such as ozone and $CO_2$ – toxic on their own, and even more so when mixed together. Several studies have explicitly examined the association between air pollution mixes and daily death rates. One recent Canadian study reported an increased risk of early death linked to a mixture of gaseous air pollution, rather than dust and particles.[49] Another, also from Canada, subsequently found a strong association between ambient concentrations of carbon monoxide and premature death across all seasons.[50]

Across the world, data from South America,[51] Europe,[52] Asia[53] and Mexico[54] all show that varying air pollutant concentrations, weather conditions and population characteristics can combine to produce adverse health effects.

Pollutants also become more toxic through interactions with local weather conditions. The combined effects of heat and pollution, for instance, cause more illness and death than either in isolation.[55] In New York City, there is a recognised relationship between acute respiratory episodes and days with high air pollution, low temperatures and high barometric pressure.[56] Other studies report that air pollution and high air temperatures in Athens raise morality,[57] while the combined effect of $N_2O$ and high temperatures have been linked to higher rates of lung cancer in Japan.[58]

As our cities expand and grow, the potential for more heatwaves and more pollution-related effects is expected to increase.[59]

## Beyond city limits

But cities also impact weather patterns outside their immediate boundaries. Research at NASA's Goddard Space Flight Centre shows that, as a result of the urban heat island effect, large coastal cities create their own rainfall and affect the weather in areas nearby.[60] The finding, based on data collected in Houston, Texas between 1998 and 2002, showed that the average rainfall rates during the warm season were 44 per cent greater downwind of Houston than upwind, even though both regions share the same climate. They also found that rainfall rates were 29 per cent greater over the city than upwind of it. Previous studies have also shown that urban heat islands create more heavy rain in, and downwind of, cities like Atlanta, St Louis and Chicago.

The warmer climate in cities does, however, mean that there are fewer days of freezing rain in this environment. US data shows that, on average, New York City has between one and two fewer days of freezing rain each year compared to the surrounding area and a shorter freezing rain season. St Louis and Washington DC also have fewer days of freezing rain.[61]

## Man changes the weather

For two million years, the earth's climate has been dominated by periodic ice ages, each lasting tens of thousands of years. These ice ages are separated by warmer 'interglacial' periods, such as the one we are in now.

During these ice ages and interglacial periods, the climate will fluctuate naturally, sometimes dramatically. Uncharacteristically, the climate of the past 9,000 years appears to have been exceptionally stable. No one knows why the expected fluctuations have not occurred during the current interglacial period, but this unexpected window of climatic stability is most likely what has allowed the development of human civilisation.[62]

To understand the significance of climate changes, it is important to distinguish between natural weather cycles (such as the changing seasons), transitory climate fluctuations (such as a temporary drought), and long-term climatic change.

Climate change means that there is a progressive variation in the average weather for a particular time of year, such as winters becoming warmer. For many years scientists believed that what we now know to be progressive climate change was a purely natural phenomenon that had nothing whatsoever to do with human intervention. We could not, we were told, prevent it or affect its natural course. For these reasons, climate change was a subject largely ignored by conventional science. Indeed, it wasn't until 1988 that the United Nations assembled a panel of scientists, economists and policy makers – known as the Intergovernmental Panel on Climate Change (IPCC) – and charged it with the task of synthesising, every five years, what the scientific community has learned about our changing climate and its impacts on people and the environment.

For the past 14 years, this panel has been investigating the causes and consequences of climate change. And although its early reports (which mostly ignored the rest of the world and focused on the 2 per cent of the globe occupied by the US) concluded that human actions had little to do with climate changes, its 2001 report,[63] which examined climate change worldwide, represented a damning and frightening about-face.

## Hot, hot, hot

The earth is getting warmer[64] and the increase in surface temperatures since 1975 has been faster than for any previous period of equal length.[65] In what can only be described as a landmark in the science of global climate change, the IPCC report confirmed, not only that global warming is occurring, but that: 'There is new and stronger evidence that most of the warming observed over the last 50 years is attributable to human activities.'

This warming trend is already upon us. Global surface temperatures in 1998 were the warmest since reliable instrumental measurements began, and seven of the ten warmest years on record occurred during the 1990s. The scientists who compiled the IPCC report project that if no major efforts are undertaken to reduce the emissions of greenhouse gases in the time period between 1990 and 2100, the earth's average surface temperature will increase between 2.5 and 10.4°F (1.4 to 5.8°C) – significantly higher than earlier predictions.

Warmer weather has lengthened the freeze-free season in many mid- and high latitude regions. The IPCC reports that the earth has lost about 10 per cent of its snow cover since the 1960s, and that lakes and rivers in the high latitudes of the Northern Hemisphere remain frozen over for two weeks less than they did a century ago. Glaciers in non-polar regions are also retreating. Arctic sea ice has thinned by some 40 per cent since the 1950s, and the surface area that it covers during the spring and summer is now 10–15 per cent smaller.

Some may react to such predication with a hearty 'So what? The temperatures where I live vary by much more than that in the course of a *day*.' However, weather changes as experienced every day are a very different matter to long-term permanent climate change. The difference might be better understood in an analogy of alcohol consumption. Most of us could drink too much every once in a while without undue effect on our health. But getting drunk every day would, without doubt, alter the way we function, lead to premature illness and possibly death. Thus it is with our planet.

Interestingly, the IPCC report also noted that certain natural phenomena also increase global warming. The scientists looked at solar activity and found that, even when taking the 11-year solar cycle into account, the sun has become more radiant during the first half of the 20th century. The weight of the evidence, however, shows unequivocally that the enhanced warming from human activities far outweighs the ups and downs in surface temperature caused by solar activity.

Some observers have tried to put a happy spin on the IPCC findings suggesting that very high temperatures in hot regions may benefit the human race. They may create high winds that at times may disperse pollution. They may reduce snail populations, which have a role in transmitting the parasitic disease schistosomiasis. Likewise, global warming might reduce the number of deaths and illnesses relating to winter weather. This is probably true, but it in addition to warmer winters global warming would also mean hotter summers and, in particular, much hotter night-time temperatures. The increased incidence of death and illness during these super-charged summers would probably negate any advantages of a warmer winter; thus most scientists predict a net disbenefit of global warming.[66]

All in all, these figures are extremely significant, revealing that rapid changes are afoot. The rate of this change, however, may ultimately depend on a highly complex and not completely understood series of 'feedback loops' whereby climate, oceans, the biosphere and the world's water cycle, including its ice sheets and glaciers, interact with each other. As with the human body, changes in any one of these important systems can affect the functioning of any of the others.[67] The relatively recent addition of human activity to this feedback loop has made the relationship that much more tricky and difficult to predict. Nevertheless, most scientists agree that global warming is likely to result in several broad trends. For instance:

- ***Sea level*** could rise by 3.5 to 34.6 inches between 1990 and 2100, making coastal groundwater saltier, endangering wetlands and inundating valuable land and coastal communities. Small island states and countries close to sea level – such as the smaller Pacific islands and places with extensive low-lying coastal areas such as Bangladesh – are especially vulnerable to sea-level rise.

- ***Precipitation*** patterns will change and this could have a significant knock-on effect, reducing rainfall in already dry regions.
- ***Plants and animals*** may not be able to adapt or migrate to new locations. Many species may die out altogether.
- ***Vegetative cover*** of the planet may change. This is relevant because trees and forests are home to many animal species. Additionally, a small change in temperature can cause huge changes in food production. Using computer-generated projections, warmer, drier weather in the Northern Hemisphere and wetter weather in tropical climates is expected to lead to lower or different crop yields, more hunger and increased disease. Scientists predict that the number of people worldwide who will be hungry will double by 2060.[68]
- ***Human health*** will suffer from increasing heat stress, worsening air pollution, declining water quality and the spread of infectious diseases. Additionally, climate changes lead to population migration, altering social groups, employment opportunity and the economic stability of many areas.

## Greenhouse gases

The earth's gaseous atmosphere functions like the panes of a greenhouse, letting some radiation from the sun in as well as retaining heat. For most of the earth's existence, this greenhouse effect has produced just enough heat to promote life and maintain equilibrium. By adding more gases to the mix, human activities have disturbed that equilibrium, effectively turning up the heat in the greenhouse.

Carbon dioxide ($CO_2$) is arguably the most important of

the greenhouse gases, and atmospheric concentrations of $CO_2$ have increased significantly since pre-industrial times. Several others gases – namely methane ($CH_4$), nitrous oxide ($N_2O$), ozone ($O_3$) and halocarbons – also contribute to the warming of the lower part of the atmosphere.

The burning of fossil fuels means that $CO_2$ levels in our atmosphere are 31 per cent higher than they were at the start of the Industrial Revolution (circa 1750). While it is true that human contributions to atmospheric $CO_2$ are small compared to those generated in nature, this seems to be missing the point. $CO_2$ generated by nature is part of a natural cycle. Human emissions are not. They are, quite literally, the straw that breaks the camel's back. Or, in scientific terms, they might be seen as part of what is called the 'butterfly effect'.

This effect was first noticed in 1961 when Edward Lorenz, a mathematician and meteorologist, was using a computer to study the way weather systems work. The first calculations on the computer provided data for short-term weather systems. But Lorenz wanted to understand how weather systems behaved over the longer term, and so re-entered the data for a second run. Unbeknownst to him, he had programmed an error into the machine by rounding off a number 0.506127 to 0.506. That tiny fraction completely changed the prediction for long-term weather patterns.

This discovery, that seemingly small intervention leads to vast, dynamic and largely unpredictable change on a global scale, he dubbed the 'butterfly effect'. According to Lorenz, 'The flapping of a single butterfly's wing today produces a tiny change in the state of the atmosphere. Over a period of time, what the atmosphere actually does diverges from what it would have done. So, in a month's time, a tornado that would have devastated the Indonesian coast doesn't happen. Or maybe one that wasn't going to happen, does.'

The butterfly effect has most recently acquired a more technical name – 'sensitive dependence on initial conditions' – but evidence of it continues to emerge in scientific literature.

## Ozone

Global warming and ozone depletion are two separate but related threats. While global warming and the greenhouse effect refer to the warming of the lower part of the atmosphere (the troposphere) due to increasing concentrations of heat-trapping gases, the ozone hole refers to the loss of ozone in the upper part of the atmosphere, the stratosphere. This is of serious concern because stratospheric ozone blocks incoming ultraviolet radiation from the sun, some of which is harmful to plants, animals and humans.

The two problems are related in a number of ways. Some human-made gases, called chlorofluorocarbons, trap heat *and* destroy the ozone layer. Currently, these gases are responsible for less than 10 per cent of total atmospheric warming, far less than the contribution from the main greenhouse gas, carbon dioxide.

The ozone layer also traps heat, so if it gets destroyed, the upper atmosphere actually cools, thereby offsetting part of the warming effect of other heat-trapping gases. But that's no reason to rejoice: trapping heat in the lower parts of the atmosphere leads to cooling of the upper part of the atmosphere. This cooling effect can produce changes in the climate that affect weather patterns in the higher latitudes. Also, the colder it gets, the greater the destruction of the protective ozone layer.

Reducing ozone-depleting gases is crucial to preventing further destruction of the ozone layer, but eliminating these gases alone will not solve the global warming problem. On the other hand, efforts to reduce all types of emissions to limit global warming will also be good for the recovery of the ozone layer.

The need to reduce industrial emissions is very acute. But even if the nations of the world stopped emitting heat-trapping gases immediately, the IPCC estimate that the climate would not stabilise for many decades because the gases we have already released into the atmosphere are long lasting – destined to stay in our atmosphere for years or even centuries. Our current 'business as usual' approach to the greenhouse effect, however, means we are at greater risk of some irreversible climate changes and the health effects associated with them.

## Weather and health revisited

Humans, because of our social organisation and cultural practices, are better protected against environmental stressors than many other plant and animal species. Hence, we are likely to feel the effects of global warming more gradually than other living organisms. But we will eventually feel them, and greenhouse gases and depletion of the ozone layer have already been likened in the medical press to an act of 'biopolitical terrorism' against the citizens of the world.[69]

Because the study of global warming is a very new science – most of which involves weather prediction models and not actual human research – it is impossible to say with any certainty what the effect on humans will be. Nevertheless, health and climate change are inextricably linked[70] and a precautionary approach may well be warranted.[71]

As most of the evidence in this book has shown, weather-related mortality increases as temperatures reach cold or hot extremes. The ideal ambient temperature for humans ranges from 3 to 25° C/38 to 77° F.[72] Through time, major heatwaves have occurred approximately every 310 years; but now scientists predict that this frequency will increase to every five to six years.[73]

Regionally and locally, a small increase in average temperature can cause relatively large increases in the number of extremely hot days, increasing the likelihood of 'killer' heatwaves during the warm season.[74] In temperate climates, the number of very hot days would approximately double for an increase of 2 to 3° C/3.6 to 5.4° F in average summer temperatures.

Increased heatwaves due to climate change would cause more heat-related illness and death. By 2020, summer mortality is expected to increase significantly (winter mortality is also expected to increase but in a less dramatic fashion).[75] It is still unclear whether the excess mortality will be offset by a decrease in deaths due to extreme cold.[76]

The combination of the sun becoming more radiant and increased UV radiation reaching the surface of earth (due to a combination of greenhouse gases and ozone depletion) may also lead to an increase in rates of skin cancer.[77] Without measures to control environmental emissions, levels are expected to quadruple by the year 2100, turning skin cancer from a relatively minor disorder resulting from our own lax habits to an important public health problem.

As average global surface temperatures gradually rise, we will likely experience more types of extreme weather.[78] This is because warmer air temperatures accelerate the water cycle – the process in which water vapour, mainly from the oceans, rises into the atmosphere before condensing and turning into precipitation (e.g. rain, fog, snow). As the oceans heat up, the rate of evaporation increases. Warm air holds more moisture than cold air and, as this extra water condenses, it is likely to drop from the sky as larger downpours.

The largest changes in precipitation are expected at mid to high latitudes.[79] For the United States, precipitation since 1970 has averaged about 5 per cent more than in the previous 70 years,[80] and cold season precipitation has increased by almost 10 per cent during the last century. Over the period 1950 to 1990, annual snowfall increased by about 20 per cent over northern Canada and by about 11 per cent over Alaska.[81] Observations for the last 100 years indicate that extreme precipitation events (more than 2 inches in 24 hours) in the United States have increased by about 20 per cent.[82]

Increased precipitation means increased risk of floods. Flooding is dangerous to human health in many ways. It can, for instance, increase populations of waterborne viruses and bacteria. Floods can wash sewage and other sources of pathogens (such as *cryptosporidium*) into supplies of drinking water. They also flush fertiliser into water supplies. Fertiliser and sewage can each combine with warmed water to trigger large-scale blooms of harmful algae. Some of these blooms are directly toxic to humans who inhale their vapours; others contaminate fish and shellfish, which, when eaten, cause illness. Algal blooms are responsible for, among other things, the transmission of *Vibrio cholerae*, the micro-organism that causes cholera. Floods also raise the risk of dengue fever and mosquito-borne infections: malaria and Rift Valley fever (a flu-like disease that can be lethal to livestock and people alike).

Climate change can extend the geographical range and reproductive season for carriers of disease such as insects, rodents and snails.[83] It may sound like a Hollywood movie, but it is already happening. In New York City, an encephalitis outbreak in the summer of 1999 claimed three lives and prompted widespread pesticide spraying. The Centers for Disease Control have identified the West Nile virus – transmitted by mosquitoes that feed on infected birds – as being responsible for this outbreak.[84] The disease, which had not been previously documented in the Western Hemisphere, occurs primarily in the late summer or early autumn in temperate regions, but can occur year round in milder climates.[85]

A recent study found a 60 per cent rise in human plague cases in New Mexico following wetter than average winter–springtime periods.[86] Plague has only been in New Mexico since the 1940s, but a large increase in reported cases that occurred in the 1970s and 1980s was associated with the wetter than normal conditions.

Likewise tickborne (viral) encephalitis in Sweden may have increased in response to a succession of warmer winters over the past two decades.[87] There is some evidence suggesting that malaria has spread to higher altitudes in the eastern African highlands in response to local warming.[88]

Parallel to the likely increase in heavy precipitation events in winter, increased temperatures will also amplify the drying out of soils and vegetation due to increased evaporation in the summer. This is likely to result in more severe and widespread droughts.

Some regional decreases in precipitation have already been observed, especially in parts of Africa, the Caribbean and tropical Asia. But certain high northern latitudes may increasingly be at risk from the effects of greenhouse gases.[89] In these regions, less winter snow and warmer temperatures lead to an earlier drying of soils in the spring, increasing the likelihood of drought.

Parching can also exaggerate the difference in atmospheric pressure between land and water leading to turbulent winds, tornadoes and other powerful storms. In addition, the altered pressure and temperature gradients that accompany global warming can cause storms, floods and droughts in areas not prone to them.

### El Niño

Future flooding and parching of the land may link into the way that global warming intensifies the El Niño Southern Oscillation (ENSO). Certain weather phenomena have global effects we are only now starting to understand. Except for the change of seasons, ENSO is now believed to have the greatest influence on the earth's weather. ENSO originates in the Pacific Ocean and was named by Peruvian and Ecuadoran fisherman who noticed periodic changes in their catch of anchovies. These tiny fish are very sensitive to water temperature changes and, as the seas warmed, the anchovy catch would decline. This warming effect peaked around Christmas, thus the fisherman gave it the name 'El Niño', Spanish for 'the Christ Child'.

Modern scientists refer to this phenomenon as ENSO because it is now known that the warming trend of El Niño is part of a cycle that also includes La Niña, a cooling trend in Pacific waters.

The fishermen's observations that El Niño produced often disastrous changes in weather have been proven by modern science.[90] On average, ENSO produces periodic short-term weather changes once every five years.[91] Specifically, it causes changes in prevailing winds that warm the ocean. As the surface temperature of the sea rises, it interacts with cooler air coming from the land to produce dramatic cold or warm fronts and

## Weathering the future

Climate change is inevitable. It is part of the natural cycle of the earth. That climate varies on a timescale of hundreds of years is demonstrated, for example, by the choice of the name 'Greenland' by the Vikings in the Medieval Warm Period. From the 11th to 13th centuries, Greenland was still 'green' due to the mild climate. Today the name seems wildly inappropriate.

Yet it is hard to look at the evidence for global warming and not ask why, with all our scientific sophistication, we have not

an increased likelihood of thunderstorms. These thunderstorms feed both moisture and wind energy into the upper atmosphere, where they influence jet stream winds, which in turn, move storms on different paths than usual, upsetting normal patterns of wet and dry weather. In some areas, El Niño conditions are twice as likely to result in drought compared to non-El Niño years.

While most of El Niño's effects are felt in the Pacific regions, it has recently been linked with changes in the sea ice as far away as Antarctica. NASA researchers report that four recent El Niño episodes over a 17-year period occurred at the same time as ice cover retreats in the Bellingshausen and Amundsen seas, suggesting that El Niño's effects are far more wide reaching than previously believed.[92]

In many locations, especially in the tropics, La Niña produces the opposite climate effects to El Niño producing drought where El Niño brought rain and vice versa.

According to the World Health Organization, the direct health impacts of ENSO are likely to include increased numbers of people made homeless due to natural disasters as well as increase in epidemic diseases from malaria to dengue and Rift Valley fever and Australian Encephalitis (Murray Valley Encephalitis, or MVE).[93]

been able to identify humans' contribution to this phenomenon and make amends much earlier. In part, it can be put down to that all-too-human blind spot, highlighted in the very first chapter of this book, that makes us want to separate ourselves from everything else that goes on in nature.

As far as technology has taken us, our bodies remain tethered to nature's rhythms. If we alter those rhythms, our health and well being will also be affected. Our ability to 'see what's around the corner' is invariably limited. But as Lorenz's butterfly effect suggests, seemingly simple systems can have astonishingly complicated dynamics.

For this reason, we should strive to gain as much insight as we can into the synchronous way in which our world and our bodies function and make better use of what ecologists call 'The Precautionary Principle' – a philosophy that acknowledges in part that science, because of its limitations and uncertainties, is poorly placed to provide an accurate picture of how humans will react to any outside influences. The principle itself is aimed mostly at the influence of toxic substances on human health, but may just as easily be applied to the effects of climate change. It suggests that our species might benefit from erring on the side of caution, and might even be best summed up in folklore:

> **For want of a nail, the shoe was lost;**
> **For want of a shoe, the horse was lost;**
> **For want of a horse, the rider was lost;**
> **For want of a rider, the battle was lost;**
> **For want of a battle, the kingdom was lost.**

# References

## Chapter 1: The Human Barometer


1 Quoted in Tromp, SW, *Biometeorology: The Impact of the Weather and Climate on Humans and Their Environment*, London: Heyden & Son Ltd, 1980.

2 Rosen, S, *Weathering: How the Atmosphere Conditions your Body, your Mind, your Moods and your Health*, New York: M Evans & Company, Inc, 1979.

3 Dobson, R, '"Health weather forecasts" to be piloted in England', *BMJ*, 2001, **322**, p 72.

4 Epstein, P, 'Climate and health', *Science*, 1999, **285**, pp 347–8.

5 Haggerty, D, *Rhymes to Predict the Weather*, Seattle, WA: Spring Meadow Publishers, 1985.

6 Landsberg, HE, *Weather and Health: An Introduction to Biometeorology*, New York: Doubleday & Company, 1969; see also Tromp, SW, 1980, *op. cit.*

7 Huntington, E, *Civilization and Climate*, 3rd edn, New Haven: Yale University Press, 1924.

8 Huntington, E, *Mainsprings of Civilization*, New York: Wiley & Sons, 1945.

9 Huntington, E, *World Power and Evolution*, 2nd edn, New Haven: Yale University Press, 1920.

10 Petersen, WF, *The Patient and the Weather*, Ann Arbor, MI: Edwards Brothers, 1935.

11 Tromp, SW, 1980, *op. cit.*

12 Tromp, SW, *Psychical Physics*, New York: Elsevier, 1949.

13 Sherretz, LA, *Weather and the Classroom Climate*, doctoral dissertation, Boulder, CO: University of Colorado, 1984; see also Landsberg, HE, 1969, *op. cit.*

14 D'Agnese, J, 'Is the weather driving you crazy?', *Discover*, 2000, **21**(6), pp 78–81. Accessed at http://ehostvgw11.epnet.com/delivery.asp?deliveryoption=Citation+with+formatted+full+text.

15 O'Connor, RP and Persinger, MA, 'Geophysical variables and behavior: LXXXII. A strong association between sudden

infant death syndrome and increments of global geomagnetic activity – possible support for the melatonin hypothesis', *Percept Mot Skills*, 1997, **84**, pp 395–402; O'Connor, RP and Persinger, MA, 'Geophysical variables and behavior: LXXXV. Sudden infant death, bands of geomagnetic activity, and pc1 (0.2 to 5 HZ) geomagnetic micropulsations', *Percept Mot Skills*, 1999, **88**, pp 391–7.

16 Coghill, RW, 'The electric railway children: an electromagnetic aetiology for cot death?', *Hosp Equip Supplies*, 1989, June, p 9.

17 D'Agnese, J, 2000, *op. cit.*

18 Reed, A, *Romantic Weather*, Hanover and London: University Press of New England, 1983.

19 Freier, GD, *Weather Proverbs*, Tuscon, AZ: Fisher Books, 1992.

20 Höppe, P *et al*, 'Prevalence of weather sensitivity in Germany', 15th Conference on Biometeorology and Aerobiology, 28 October–1 November 2002 in Kansas City, MO.

21 Cited in Kaiser, M, *How the Weather Affects your Health*, Melbourne: Michelle Anderson, 2002.

22 Reported in Landsberg, HE, 'Weather, climate and you', *Weatherwise*, 1986, Oct, p 248.

23 Reported in Henson, R, 'Smells like rain', *Weatherwise*, 1996, Apr/May, p 29.

24 Persinger, MA, 'Lag responses in mood, reports to changes in the weather matrix', *International Journal of Biometerology*, 1975, **19**, pp 108–114.

25 Kalkstein, LS and Valimont, KM, 'Climate effects on human health', *Potential Effects Of Future Climate Changes Of Forests And Vegetation Agriculture, Water Resources, And Human Health*. EPA Science and Advisory Committee, Monograph no 25389. Washington, D.C.: US Environmental Protection Agency, **12**, 1987, pp 122–52.

26 Rosen, S, 1979, *op. cit.*

27 Parrish, Gib, 'Impact Of Weather On Health', Atlanta: Centers for Disease Control, 1999.

28 Avrach, WW, 'Climatotherapy at the Dead Sea', in Farber, EM (eds), *Proceedings of the Second International Symposium, 1976*, New York: York Medical Books, 1977.

## Chapter 2: Of Seasons and Cycles

1 Petersen, WF, Man, *Weather, Sun, Springfield*, Illinois: Charles C Thomas, 1947.

2 Scheving, LE and Halberg, F, *Chronobiology: Principles and Applications to Shifts in Schedules*, Boston: Kluwer Academic, 1981.

3 Axt, A, 'Autism viewed as a consequence of pineal gland malfunction', *Farmakoterapia w Psychiatrii i Neurologii*, 1998, **98**, pp 112–134.

4 Halberg, F *et al*, 'Physics of auroral phenomena', *Proc XXV Annual Seminar*, Apatity, Kola Science Center, Russian Academy of Science, 2002, pp 161–165.

5 Tarquini, B *et al*, 'Chronome assessment of circulating melatonin in humans', *In Vivo*, 1997, **11**, pp 473–84.

6 Carskadon, MA *et al*, 'Adolescent sleep patterns, circadian timing and sleepiness at a transition to early school days', *Sleep*, 1998, **21**, pp 871–81.

7 Dement, C and Vaughan, C, *The Promise of Sleep*, New York: Delacorte Press, 1999.

8 Carrier, KJ and Monk, TH, 'Effects of sleep and circadian rhythms on performance', in Turek, FW and Zee, PC (eds), *Regulation of Sleep And Circadian Rhythms*, New York: Dekker Inc, 1999.

9 Blake, MJF, 'Temperament and time of day', in Colquhoun, WP (ed), *Biological Rhythms And Human Performance*, London: Academic Press, 1971; Monk TH, 'What can the chronobiologist do to help the shift worker?', *J Biol Rhythms*, 2000, **15**, pp 86–94.

10 Reilly, TG *et al*, *Biological Rhythms and Exercise*, Oxford: Oxford University Press, 1997.

11 Broughton, R, 'Biorhythmic variations in consciousness and psychological functions', *Can Psychol Rev*, 1975, **16**, pp 217–39.

12 Hayashi, M *et al*, 'Ultradian rhythms in task performance, self-evaluation, and EEG activity', *Percept Motor Skill*, 1994, **29**, pp 791–800; Aeschbach, D *et al*, 'Two circadian rhythms in the human electroencephalogram during wakefulness', *Am J Physiol*, 1999, **277**, pp R1771–9.

13 Scheving, LE, 'Mitotic activity in the human epidermis', *Anatom Rec*, 1959, **135**, pp 7–20.

14 Reinberg, A *et al*, 'Oral contraceptives alter circadian rhythm parameters of cortisol, melatonin, blood flow, transepidermal water loss and skin amino acids of healthy young women', *Chronobiol Int*, 1996, **13**, pp 199–211.

15 Verschoore, M *et al*, 'Circadian rhythms in the number of actively secreting sebaceous follicles and androgen circadian rhythms', *Chronobiol Int*, 1993, **5**, pp 349–59; Yosipovitch, G

*et al*, 'Time-dependent variations of the skin barrier function in humans: transdermal water loss, stratum corneum hydra tion, skin surface pH, and skin temperature', *J Invest Dermatol*, 1998, **110**, pp 20–23.

16 Scheving, LE and Halberg, 1981, *op. cit.*

17 Moore, JG and Goo, RH, 'Day and night aspirin induced gastric mucosal damage and protction by ranitidine in man', *Chronobiol Int*, 1987, **4,** pp 111–16.

18 Levi, FC *et al*, 'Chronotherapy of osteoarthritic patients: optima tion of indomethacin sustain release (ISR)', *Ann Rev Chronopharmacol*, 1984, **1,** pp 345–8.

19 Rejholec, VV *et al*, 'Preliminary observations from a double blind crossover study to evaluate the efficacy of flurbiprofen given at different times of day in the treatment of rheumatoid arthritis', *Ann Rev Chronopharmacol*, 1984, **1**, pp 357–60.

20 Kowanko, IC *et al*, 'Circadian variations in the signs and symptoms of rheumatoid arthritis and in the therapeutic effectiveness of flurbiprofen at different times of the day', *Br J Clin Pharmacol*, 1981, **11**, pp 477–84; Kowanko, IC *et al*, 'Domiciliary self-measurements of in rheumatoid and the demonstration of circadian rhythmicity', *Ann Rheum Dis*, 1982, **41**, pp 453–5; Swannell, AJ, 'Biological rhythms and their effect in the assesment of disease activity in rheumatoid arthritis', *Br J Clin Pract*, 1983, **38** (suppl 33), pp 16–19; Hark ness, JAL *et al*, 'Circadian variation in disease activity in rheumatoid arthritis', *BMJ*, 1982, **284**, pp 551–4.

21 Bouchoucha, M *et al*, 'Day–night patterns of gastroesophagal reflux', *Chronobiol Int*, 1995, **12**, pp 267–77.

22 Moore, JG and Halberg, F, 'Circadian rhythm of gastric acid secretion in active duodenal ulcer: chronobiological statistical characteristics and comparison of acid secretory and plasma gastrin patterns with healthy subjects and post-vagotomy and pyloroplasty patients', *Chronobiol Int*, 1987, **4**, pp 101–10.

23 Hatlebakk, JG *et al*, 'Nocturnal gastric acidity and acid break through on different regimens of omeprazole 40 mg daily', *Aliment Pharmacol Therap*, 1998, **12**, pp1235–40; Ireland, A *et al*, 'Ranitidine 150 mg twice daily vs. 300 mg nightly in treat ment of duodenal ulcers', *Lancet*, 1984, **ii**, pp 274–6; see also Howden, CW and Hunt, RH, 'The histamine H2-receptor antagonists, in Swabb, EA and Szabo, S (eds), *Ulcer Disease: Investigation and Basis for Therapy*, New York: Marcel Dekker, 1991; Merki, H *et al*, 'Single dose treatment with H2-receptor

antagonists: is bedtime administration too late?' *Gut*, 1987, **28**, pp 451–4.

24 Haus, E *et al*, 'Introduction to chronobiology', in Sheving, LE and Halberg, F (eds), *Chronobiology: Principles and Applications to Shifts in Schedules*, Netherlands: Sijthoff & Noorhoff International Publishers, 1980; Hyman, JW, *The Light Book: How Natural and Artificial Light Affect our Health, Mood and Behavior*, Los Angeles: Jeremy P Tarcher, 1990.

25 Cited in Halberg, F *et al*, 'The story behind: pineal mythology and chronorisk. The swan song of Brunetto Tarquini', *Neuroendocrinol Lett*, 1999, **20**, pp 91–100.

26 Ursin, G *et al*, 'Mammographic density changes during the menstrual cycle', *Cancer Epidemiol Biomarkers Prev*, 2001, **10**, pp 141–2; White, E *et al*, 'Variation in mammographic breast density by time in menstrual cycle among women aged 40–49 years', *J Natl Cancer Inst*, 1998, **90**, pp 906–10.

27 Senie, RT *et al*, 'Timing of breast cancer excision during the menstrual cycle influences duration of disease-free survival', *Ann Intern Med*, 1991, **115**, pp 337–42; Badwe, RA *et al*, 'Timing of surgery during menstrual cycle and survival of premenopausal women with operable breast cancer', *Lancet*, 1991, **337**, pp 1261–4; Saad, Z *et al*, 'Timing of surgery in relation to the menstrual cycle in premenopausal women with operable breast cancer', *Br J Surg*, 1994, **81**, pp 217–20.

28 Wehr, TA, 'Melatonin and seasonal rhythms', *J Biol Rhythms*, 1997, **12**, pp 518–27; Wehr, TA, 'The durations of human melatonin secretion and sleep respond to changes in daylength (photoperiod)', *J Clin Endocrinol Metab*, 1991, **73**, pp 1276–80.

29 Lewy, AJ *et al*, 'Morning vs evening light treatment of patients with winter depression', *Arch Gen Psychiatr*, 1998, **55**, pp 890–6.

30 Sack, RL *et al*, 'Human melatonin production decreases with age', *J Pineal Res*, 1986, **3**, pp 379–88.

31 Tarquini, B *et al*, 'Circadian mesor-hyperprolactinemia in fibrocystic mastopathy', *Am J Med*, 1979, **66**, pp 229–37.

32 Bulbrook, M *et al*, 'Metachronanalyses of prolactin (prl) and human breast (B) cancer', *Chronobiologia*, 1987, **14**, p 156.

33 Wehr, TA, 'Photoperiodism in humans and other primates, evidence and implications', *Biol Rhythms*, 2001, **16**, pp 348–64.

34 Wehr, TA *et al*, 'Suppression of men's responses to seasonal changes in day length by modern artificial lighting', *Am J*

*Physiol*, 1995, **269** (1 Pt 2), pp R173–8.

35 Aschoff, J, 'Annual rhythms in man', in Aschoff, J (ed), *Handbook of Behavioral Neurobiology*, New York: Plenum, 1981.

36 Rosenthal, NE and Wehr, TA, 'Seasonal affective disorders', *Psychiatr Ann*, 1987, **17**, pp 670–4.

37 Tromp, SW, 'Influence of weather and climate on the fibrinogen content of human blood', *Int J Biometeorol*, 1972, **16**, pp 93–5; Bull, GM *et al*, 'Relationship of air temperature to various chemical, haematological and haemostatic variables', *J Clin Pathol*, 1979, **32**, pp 16–20.

38 Doupe, D *et al*, 'Seasonal fluctuations in blood volume', *Can J Biochem Physiol*, 1957, **35**, pp 203–13.

39 Rose, G, 'Seasonal variation in blood pressure in man', *Nature*, 1961, **189**, p 235.

40 Carlson, LD and Hsieh, ACL, 'Cold', in Edholm, OG (ed), *The Physiology of Human Survival*, New York: Academic Press, 1965; Matsui, H *et al*, 'Seasonal variation of aerobic work capacity in ambient and constant temperature', in Folinsbee, LJ (ed), *Environmental Stress, Individual Human Adaptations*, New York: Academic Press, 1978.

41 Billewicz, WL, 'A note on body weight measurements and seasonal variation', *Human Biol*, 1967, **39**, pp 242–50.

42 Caplan, CE, 'The big chill, diseases exacerbated by exposure to cold', *CMAJ*, 1999, **160**, p 33.

43 Stout, RW *et al*, 'Seasonal changes in haemostatic factors in young and elderly subjects', *Age Aging*, 1996, **25**, pp 256–9.

44 Smith, R, 'Doctors and climate change, action is needed because of the high probability of serious harm to health', *BMJ*, 1994, **309**, pp 1384–6; Blindauer, KM *et al*, 'The 1996 New York blizzard, impact on non-injury emergency visits', *Am J Emerg Med*, 1999, **17**, pp 23–7.

45 Rietveld, WJ *et al*, 'Seasonal fluctuations in the cervical smear rates for (pre)malignant changes and for infections', *Diagnost Cytopathol*, 1997, **17**, pp 452–5; Hermida, RC and Ayala, DE 'Reproducible and predictable yearly pattern in the incidence of uterine cervical cancer', *Chronobiol Int*, 1996, **13**, pp 305–16.

46 Levi, FA *et al*, 'Seasonal modulation of the circadian time structure of circulating T and natural killer lymphocyte subsets from healthy subjects', *J Clin Invest*, 1998, **81**, pp 407–13.

47 Cowgill, UM, 'Season of birth in man. Contemporary situation with special reference to Europe and the Southern Hemisphere', *Ecology*, 1966, **47**, pp 614–23; Calot, G and

Blayo, C, 'Recent course of fertility in Western Europe', *Population Studies*, 1982, **36**, pp 345–72.

48 Macfarlane, WV, 'Seasonality of conception in human populations', *Int J Biometeorol*, 1970, **14** (Suppl 4,1), pp 167–82.

49 Seiver, DA, 'Trend and variation in the seasonality of US fertility, 1947–1976', *Demography*, 1985, **22**, pp 89–100.

50 Shimura, M *et al*, 'Geographical and secular changes in the seasonal distribution of births', *Soc Sci Med*, 1981, **15D**, pp 103–109.

51 Bernard, RP *et al*, 'Seasonality of birth in India', *J Biosocial Sci*, 1978, **10**, pp 409–21.

52 Becker, S, 'Seasonality of fertility in Matlab, Bangladesh', *J Biosocial Sci*, 1981, **13**, pp 97–105.

53 Johnston, JE and Branton, C, 'Effects of seasonal climatic changes on certain physiological reactions, semen production and fertility of dairy bulls', *J Dairy Sci*, 1953, **36**, pp 934–42; Glover, TD, 'The effect of scrotal insulation and the influence of the breeding season upon fructose concentration in the semen of ram', *J Endocrinol*, 1956, **13**, pp 235–42; Venkatachalam, PS and Ramanathan, KS, 'Effects of moderate heat on the testes of rats and monkeys', *J Reprod Fertil*, 1962, **4**, pp 51–6.

54 Tjoa, WS *et al*, 'Circannual rhythm in human sperm count revealed by serially independent sampling', *Fertil Steril*, 1982, **38**, pp 454–9.

55 McKeown, T and Record, RG, 'Seasonal incidence of congenital malformations of the central nervous system', *Lancet*, 1951, **1**, pp 192–6; Wehrung, DA and Hay, S, 'A study of seasonal incidence of congenital malformations in the United States', *Br J Prev Soc Med*, 1970, **24**, pp 24–32; Cohen, P, 'Seasonal variations of congenital dislocation of the hip', *J Interdisciplinary Cycle Res*, 1971, **2**, pp 417–25.

56 Slater, BCS *et al*, 'Seasonal variation in congenital abnormalities, preliminary report of a survey conducted by the Research Committee of Council of the College of General Practitioners', *Br J Prev Soc Med*, 1964, **18**, pp 1–7.

57 Elwood, JH and MacKenzie, G, 'Comparisons of secular and seasonal variations in the incidence of anencephalus in Belfast and four Scottish cities, 1956–1966', *Br J Prev Soc Med*, 1971, **25**, pp 17–25.

58 Torrey, EE *et al*, 'Seasonality of schizophrenic births in the United States', *Arch Gen Psychiatry*, 1977, **34**, pp 1065–70; Pulver, AE *et al*, 'The association between season of birth and the risk for

schizophrenia', *Am J Epidemiol*, 1981, **114**, pp 735–49.
59 Vaiserman, AM *et al*, 'Seasonal programming of adult longevity in Ukraine', *Int J Biometeorol*, 2002, **47**, pp 49–52.

## Chapter 3: The Sun and the Moon

1 Hillman, D *et al*, 'About yearly (circadecennian) cosmo-heliogeomagnetic signature in Acetabularia', *Scripta Med (Brno)*, 2002, **75**, pp 303–8.
2 Stoupel E *et al*, 'Intraocular pressure (IOP) in relation to four levels of daily geomagnetic and extreme yearly solar activity', *Int J Biometeorol*, 1993, **37**, pp 42–5.
3 Allahverdiyev, AR *et al*, 'Possible space weather influence on functional activity of the human brain', *Proceedings of the Space Weather Workshop, Looking Towards a European Space Weather Programme*, December 2001.
4 Caniggia, M and Scala, C, 'Sunspots and hip fractures', *Chronobiologia*, 1991, **18**, pp 1–8.
5 Anonymous, 'Solar activity and terrestrial thunderstorms', *New Scientist*, 1979, **81**, p 256.
6 Halberg, F *et al*, 'Cross-spectrally coherent 10.5- and 21-year biological and physical cycles, magnetic storms and myocardial infarctions', *Neuroendocrinol Lett*, 2000, **21**, pp 233–58; Baevsky, RM *et al*, 'Meta-analyzed heart rate variability, exposure to geomagnetic storms, and the risk of ischemic heart disease', *Scripta Med (Brno)*, 1997, **70**, pp 201–6.
7 Sitar, J, 'The causality of lunar changes on cardiovascular mortality', *Cas Lek Cesk*, 1990, **129**, pp 1425–30.
8 Petro, VM *et al*, 'An influence of changes of magnetic field on the earth on the functional state of humans in the conditions of space mission', *Proceedings of the International Symposium Computer Electro-Cardiograph on Boundary of Centuries*, Moscow, Russian Federation, 27–30 April 1999.
9 Cornelissen, G *et al*, 'Chronomes, time structures, for chronobioengineering for "a full life"', *Biomed Instrum Technol*, 1999, **33**, pp 152–87.
10 Hope-Simpson, RE, 'Sunspots and flu, a correlation', *Nature*, 1978, **275**, p 86; Hoyle, F and Wickramasinghe, NC, 'Sunspots and influenza', *Nature*, 1990, **343**, p 304.
11 Freitas, RA Jr, 'Sunspots and disease', *Omni*, 1984, **6**, p 40.

12 Radin, D, *The Conscious Universe*, San Francisco: HarperEdge, 1997.
13 Becker, R, *The Body Electric, Electromagnetism and the Foundation of Life*, New York: Quill, 1985; Raps, A *et al*, 'Geophysical variables and behavior, LXIX. Solar activity and Admission of Psychiatric Patients', *Percept Motor Skills*, 1992, **74**, pp 449–50; Reitz, G, 'Biological effects of space radiation', *Proceedings of the Space Weather Workshop, Looking towards a European Space Programme*, December 2001.
14 Friedman, H *et al*, 'Geomagnetic parameters and psychiatric hospital admissions', *Nature*, 1963, **200**, pp 626–8.
15 Allahverdiyev, AR *et al*, 2001, *op. cit.*
16 Randall, W and Moos, WS, 'The 11–year cycle in human births', *Int J Biometeorol*, 1993, **37**, pp 72–7.
17 Juckett, DA and Rosenberg, B, 'Correlation of human longevity oscillations with sunspot cycles', *Rad Res*, 1993, **133**, pp 312–20.
18 Tromp, SW, *Biometeorology: The Impact of the Weather and Climate on Humans and their Environment,* London: Heyden & Son Ltd, 1980.
19 Payne, B, 'Cycles of peace, sunspots, and geomagnetic activity', *Cycles*, 1984, **35**, p 101.
20 'Moonstruck Scientists', *Time*, 1972, 10 January.
21 Pasichnyk, RM, *The Vital Vastness – The Living Cos*mos, Writers Showcase Press, 2002.
22 Kelly, IW *et al*, 'Geophysical variables and behavior, LXIV. An empirical investigation of the relationship between worldwide automobile traffic disasters and lunar cycles, No Relationship', *Psychological Rep*, 1990, **67**, pp 987–94.
23 Raison, CL *et al*, 'The moon and madness reconsidered', *J Affect Disord*, 1999, **53**, pp 99–106.
24 Andrews, EJ, 'Moon talk, The cyclic periodicity of postoperative haemorrhage', *J Florida Med Assoc*, 1960, **46**, pp 1362–66.
25 de Castro, JM and Pearcey, SM, 'Lunar rhythms of the meal and alcohol intake of humans', *Physiol Behav*, 1995, **57**, pp 439–44.
26 Payne, SR *et al*, 'Urinary retention and the lunisolar cycle, is it a lunatic phenomenon?', *BMJ*, 1989, **299**, pp 1560–2.
27 Weigert, M *et al*, 'Do lunar cycles influence in vitro fertilization results?' *J Assist Reprod Genet*, 2002, **19**, pp 539–40.
28 Thery, A, 'The influence of the moon on bleeding', *Hist Med Vet*, 2003, **28**, pp 27–32.
29 Mikulecky, M and Rovensky, J, 'Gout attacks and lunar cycle', *Med Hypotheses*, 2000, **55**, pp 24–5.

30 Lieber, AL and Sherin, CR, 'Homicides and the lunar cycle; toward a theory of lunar influence on human emotional disturbance', *Am J Psychiatry*, 1972, **129**, pp 69–74; see also Lieber, AL, *How the Moon Affects You*, Mamaroneck, NY: Hastings House, 1996.

31 Kelly, IW *et al*, 'The moon was full and nothing happened; a review of studies on the moon and human behavior and belief', in Nickell, J *et al* (eds), *The Outer Edge*, Amherst, NY: CSICOP, 1996.

32 Thakur, CP and Sharma, D, 'Full moon and crime', *BMJ*, 1984, **289**, pp 1789–91; Soyman, P and Holdstock, TL, 'The influence of the sun, moon, climate and economic conditions on crisis incidence', *J Clin Psychol*, 1980, **36**, pp 884–93; Sitar, J, 'Chronobiology of human aggression', *Cas Lek Cesk*, 1997, **136**, pp 174–80; Ghiandoni, G *et al*, 'Incidence of lunar postion in the distribution of deliveries. A statistical analysis', *Minerva Ginecol*, 1997, **49**, pp 91–4; Sitar, J, 'The effect of the semilunar phase on an increase in traffic accidents', *Cas Lek Cesk*, 1994, **133**, pp 596–8.

33 Owen, C *et al*, 'Lunar cycles and violent behaviour', *Aust N Z J Psychiatry*, 1998, **23**, pp 496–9; DeCastro, JM and Pearcey, SM, 'Lunar rhythms of the meal and alcohol intake of humans', *Physiol Behav*, 1995, **57**, pp 439–44; Sharfman, M, 'Drug overdose and the full moon', *Percept Mot Skills*, 1980, **50**, pp 124–6; Laverty, WH and Kelly, IW, 'Cyclical calendar and lunar patterns in automobile property accidents and injury accidents', *Percept Mot Skills*, 1998, **86**, pp 299–302; Thompson, DA and Adams, SL, 'The full moon and ED patient volumes, unearthing a myth', *Am J Emerg Med*, 1996, **14**, pp 161–4.

34 Alonso, Y, 'Geophysical variables and behavior, LXXII. Barometric pressure, lunar cycle, and traffic accidents', *Percept Mot Skills*, 1993, **77**, pp 371–6.

35 Laverty, WH and Kelly, IW, 'Cyclical calendar and lunar patterns in automobile property accidents and injury accidents', *Percept Mot Skills*, 1998, **86**, pp 299–302; see also Lieber, AL, 1978, *op. cit.*

36 Watson, L, *Supernature*, London: Hodder and Stoughton, 1971.

37 Thakur, CP *et al*, 'Full moon and poisoning', *BMJ*, 1980, **281**, p 1684.

38 Jones, PK and Jones, SL, 'Lunar association with suicide', *Suicide Life Threat Behav*, 1977, **7**, pp 31–9; Gutierrez-Garcia, JM and

Tusell, F, 'Suicides and the lunar cycle', *Psychol Rep*, 1997, **80**, pp 243–50; Jacobsen, D *et al*, 'Self-poisoning and moon phases in Oslo', *Hum Toxicol*, 1986, **5**, pp 51–2; Maldonado, G and Kraus, JF, 'Variation in suicide occurrence by time of day, day of the week, month, and lunar phase', *Suicide Life Threat Behav*, 1991, **21**, pp 174–87.

39 Bhattacharjee, C *et al*, 'Do animals bite more during a full moon? Retrospective observational analysis', *BMJ*, 2000, **321**, pp 1559–61.

40 Chapman, S and Morrell, S, 'Barking mad? Another lunatic hypothesis bites the dust', *BMJ*, 2000, **321**, pp 1561–3.

41 Macdonald, L *et al*, 'Effect of the moon on general practitioners' on call workload', *J Epidemiol Community Health*, 1994, **48**, pp 323–4.

42 DeVoge, SD and Mikawa, JK, 'Moon phases and crisis calls, a spurious relationship', *Psychol Rep*, 1977, **40**, pp 387–90.

43 Oderda, GM and Klein-Schwartz, W, 'Lunar cycle and poison center calls', *J Toxicol Clin Toxicol*, 1983, **20**, pp 487–95.

44 Coates, W *et al*, 'Trauma and the full moon, a waning theory', *Ann Emerg Med*, 1989, **18**, pp 763–5; Michelson, L *et al*, 'Investigation of periodicity in crisis intervention calls over an eight-year span', *Psychol Rep*, 1979, **45**, pp 420–2; Gorvin, JJ and Roberts, MS, 'Lunar phases and psychiatric hospital admissions', *Psychol Rep*, 1994, **75**, pp 1435–40; see also Oderda, GM *et al*, 1983, *op. cit.*

45 Lieber, AL, 'Human aggression and the lunar synodic cycle', *J Clin Psychiatry*, 1978, **39**, pp 385–92.

46 Tasso, J and Miller, E, 'The effects of the full moon on human behaviour', *J Psychol*, 1976, **93**, pp 81–3.

47 Thakur, CP and Sharma, D, 'Full moon and crime', *BMJ* (Clin Res Ed), 1984, **289**, pp 1789–91.

48 Bamber, D, 'Violent moods rise with the new moon', *Sunday Telegraph*, 1998, 29 November, p 19.

49 Forbes, GB and Lebo, GR Jr, 'Antisocial behavior and lunar activity, a failure to validate the lunacy myth', *Psychol Rep*, 1977, **40** (3 Pt. 2), pp 1309–10; Owen, C *et al*, 'Lunar cycles and violent behaviour', *Aust N Z J Psychiatry*, 1998, **32**, pp 496–9; Cohen-Mansfield, J *et al*, 'Full moon, does it influence agitated nursing home residents?', *J Clin Psychol*, 1989, **45**, pp 611–4; Simon, A, 'Aggression in a prison setting as a function of lunar phases', *Psychol Rep*, 1998, **82** (3 Pt 1), pp 747–52.

50 Climent, CE and Plutchik, 'Lunar madness, an empirical study', *Compr Psychiatry*, 1977, **18**, pp 369–74.
51 Lieber, AL, 1978, *op. cit.*
52 Barr, W, 'Lunacy revisited. The influence of the moon on mental health and quality of life', *J Psychosoc Nurs Ment Health Serv*, 2000, **38**, pp 28–35.
53 Wilkinson, G *et al*, 'Lunar cycle and consultations for anxiety and depression in general practice', *Int J Soc Psychiatry*, 1997, **43**, pp 29–34.
54 Criss, TB and Marcum, JP, 'A lunar effect on fertility', *Soc Biol*, 1981, **28**, pp 75–80.
55 Menaker, W and Menaker, A, 'Lunar periodocity in human reproduction; a likely unit of biological time', *Am J Obstet Gynecol*, 1959, **77**, pp 905–14.
56 Guillon, P *et al*, 'Seasonal, weekly and lunar cycles of birth. Statistical study of 12,035,680 births', *Rev Fr Gynecol Obstet*, 1988, **83**, pp 703–8.
57 Martens, R *et al*, 'Lunar phase and birth rate; a fifty-year critical review', *Psychological Reports*, 1988, **63**, pp 923–34; Kelly, I and Martens, R, 'Lunar phase and birthrate; an update', *Psychological Reports*, 1994, **75**, pp 507–11; Periti, E and Biagiotti, R, 'Lunar phases and incidence of spontaneous deliveries. Our experience', *Minerva Ginecol*, 1994, **46**, pp 429–33.
58 Currie, RG, 'Variance contribution of Mn (sic) and Sc signals to Nile River data over (sic) a 30–8 year (sic) bandwidth', in Finkl, CW Jr (ed), 'Holocene cycles, climate, sea levels and sedimentation; a jubilee volume in celebration of the 80th birthday of Rhodes W Fairbridge', *J Coastal Res*, 1995, **17** (special issue), p 402.
59 Carpenter, T *et al*, 'Observed relationships between lunar tidal cycles and formation of hurricanes and tropical storms', *Monthly Weather Rev*, 1972, **100**, pp 451–60.
60 Wood, FJ, 'The strategic role of perigean spring tides in nautical history and North American coastal flooding 1635–1976', Washington DC: US Department of Commerce National Oceanic and Atmospheric Administration, US Government Printing Office, Stock No. 003–017–00420–1, 1978; Wood, FJ, *Tidal Dynamics, Coastal Flooding and Cycles of Gravitational Force*, Dordrecht, The Netherlands: D Reidel Publishing Company, 1985.
61 Lethbridge, M, *Solar–Lunar Variables, Thunderstorms and Tornadoes*,

Department of Meteorology Report, College of Earth and Mineral Sciences, Pennsylvania State University, University Park, 1969; Markson, R, 'Considerations regarding solar and lunar modulation of geophysical parameters, atmospheric electricity and thunderstorms', *Pure Appl Physics*, 1971, **84**, pp 61–200.

62 O'Mahoney, G, 'Rainfall and moon phase', *Quart J Royal Meteor Soc*, 1965, **91**, pp 196–208.

63 Bradley, DA *et al*, 'Lunar synodical period and widespread precipitation', *Science*, 1962, **137**, pp 748–749.

64 Brier, GW and Bradley, DA, 'Lunar synodical period and precipitation in the United States', *J Atmos Sci*, 1964, **21**, pp 386–95.

65 Berson, FA and Deacon, EL, 'Heavy rainfalls and the lunar cycle', *Indian J Meteor Geophys*, 1965, **16**, pp 5–60.

66 Kaye, CA and Stuckey, GW, 'Nodal tidal cycle of 18.6 yr', *Geology*, 1973, **1**, pp141–44.

67 Currie, RG, 'Examples and implications of 18.6- and 11-yr terms in world weather records', in Rampino, MR *et al* (eds), *Climate, History, Periodicity and Predictability, International Symposium held at Barnard College, Columbia University, New York, New York, 21–23 May 1984 Proceedings*, New York, NY: Van Nostrand Reinhold Company, 1987; Currie, RG 'Luni–solar 18.6- and 10–11-year solar cycle (sic) signals in HH Lamb's dust veil (sic) index', *J Climatol*, 1994, **14**, pp 215–26; Currie, RG, *et al* 'Deterministic signals in European fish catches, wine harvests, sea level, and further experiments', *J Climatol*, 1993, **8**, pp 255–81.

68 Olcese, J *et al*, 'Responses of the mammalian retina to experimental alteration of the ambient magnetic field', *Brain Res*, 1988, **448**, pp 325–30.

## Chapter 4: When the Wind Blows

1 Watson, L, *Heaven's Breath: A Natural History of the Wind*, London: Sceptre, 1988.

2 Durschmied, E, *The Weather Factor*, London: Coronet, 2000.

3 Kaiser, M, *How the Weather Affects Your Health*, Melbourne: Michelle Anderson, 2002.

4 Palmer, B, *Body Weather*, Harrisburg, PA: Stackpole Books, 1976.

5 Sulman, FG, *The Effect of Air Ionisation, Electric Fields, Atmospherics and other Electric Phenomena on Man and Animal*, Springfield, III: Charles C Thomas, 1980.

6 Krueger, AP, 'Are air ions biologically significant? A review of a controversial subject', *Int J. Biometeorol*, 1972, **16**, pp 313–22; Ryushi, T *et al*, 'The effect of exposure to negative air ions on the recovery of physiological responses after moderate endurance exercise', *Int J Biometeorol*, 1998, **41**, pp 1320–6.

7 Krueger, AP, 'The biological effects of air ions', *Int J Biometeorol*, 1985, **29**, pp 205–6.

8 Marin, V *et al*, 'Effects of ionization of the air on some bacterial strains', *Ann Ig*, 1989, **1**, pp 1491–1500.

9 Krueger, AP and Sobel, DS, 'Air ions and health', in Sobel, DS (ed), *Ways of Health – Holistic Approaches to Ancient and Contemporary Medicine*, New York: Harcourt Brace Jovanovich Inc, 1979.

10 Gualtierotti, R *et al*, *Bioclimatology and Aeroionotherapy*, Milan, Italy: Carlo Erba Foundation, 1968.

11 Winsor, T and Beckett, J, 'Biological effects of ionized air in man', *Am J Psychical Med*, 1958, **37**, pp 83–9.

12 Livanova, LM *et al*, 'Effect of the short–term exposure to negative air ions on individuals with autonomic disorders', Zh Vyssh Nerv Deya, 1999, **49**, pp 760–7.

13 Marin, V *et al*, 'Effects of ionization of the air on some bacterial strains', *Ann Ig*, 1989, **1**, pp 1491–500.

14 Gabbay, J *et al*, 'Effect of ionization on microbial air pollution in the dental clinic', *Environ Res*, 1990, **52**, pp 99–106.

15 Hawkins, LH, 'The possible benefits of negative-ion generators', in Pearse, BG (ed), *Health Hazards of VDTs?*, Chichester: John Wiley & Sons, 1984.

16 Terman, M and Terman, JS, 'Treatment of seasonal affective disorder with a high-output negative ionizer', *J Altern Complement Med*, 1995, **1**, pp 87–92; Terman, M *et al*, 'A controlled trial of timed bright light and negative air ionisation for treatment of winter depression', *Arch Gen Psychia*, 1988, **55**, pp 875–82.

17 Brown, GC and Kirk, RE, 'Geophysical variables and behavior, XXXVIII. Effects of ionized air on the performance of a vigilance task', *Percept Mot Skills*, 1987, **64** (3 Pt 1), pp 951–62; Nakane, H *et al*, 'Effect of negative air ions on computer operation, anxiety and salivarychromogranin A-like immunoreactivity', *Int J Psychophysiol*, 2002, **46**, pp 85–9.

18 Kornblueh, IH, 'Artificial ionization of the air and its biological significance', *Clin Med*, 1973, **69**, pp 282–286.

19 Kellogg, EW *et al*, 'Long-term biological effects of air ions and

D.C. electric fields on NAMRU mice, first year report', *Int J Biometeorol*, 1985, **29**, pp 253–68; Krueger, AP and Reed, EJ, 'Biological impact of small air ions', *Science*, 1976, **24**, pp 1209–13; Krueger, AP *et al*, 'Further observations on the effect of air ions on influenza in the mouse', *Int J Biometeorol*, 1974, **18**, pp 46–56.

20 Krueger, AP, 'The biological mechanism of air ion action', *Int J Biometeorol*, 1963, **7**, pp 3–16; Krueger, AP *et al*, 'Effects of inhaling non–ionized or positive ionized air on the blood levels of 5–HT in mice', *Int J Biometeorol*, 1966, **10**, pp 17–28.

21 Sigel, S, 'Biopsychological influences of air ions on men, effects on 5–HT and mood', *Dissertation Abstracts Int*, 1970, **40**, p 1416B; Deleanu, M, 'Donnes relatifes a l'effet normalisateur des aeroions negatifs sur la tension arterielle' in *Proceedings of the Sixth International Biometeorological Congress*, 1972, pp 46–7; Tal, E *et al*, 'Effect of ionization on blood serotonin in vitro', *Experientia* (Basel), 1976, **32**, pp 326–7.

22 Sulman, FG, 'Meteorologische frontverschiebung und wetter fuehligkeit – Foehn, Chamssin, Scharaw', *Arztliche Praxis*, 1971, **23**, pp 998–99, quoted in Sulman, FG, *Health Weather and Climate*, Basel, New York: Karger, 1976.

23 Faust, V J, *Interdiscpl Cycle Res*, 1974, **5**, p 313, quoted in Sulman, FG, *Health, Weather and Climate*, Basel, New York: Karger, 1976.

24 Sulman, FG, 'The impact of weather on human health', *Rev Environ Health*, 1984, **4**, pp 83–119; Sulman, FG, 'Migraine and headache due to weather and allied causes and its specific treatment', *Ups J Med Sci Supple*, 1980, **31**, pp 41–4; Sulman, FG *et al*, 'New methods in the treatment of weather sensitivity', *Fortschr Med*, 1977, **95**, pp 746–52.

25 Giannini, AJ *et al*, 'The serotonin irritation syndrome – a new clinical entity?', *J Clin Psychi*, 1986, **47**, pp 22–25; Giannini, AJ *et al*, 'Reversibility of serotonin irritation syndrome with atmospheric anions', *J Clin Psychi*, 1986, **47**, pp 141–143.

26 Robinson, N and Dirnfield, FS, 'The ionisation state of the atmosphere as a function of meteorological elements and the various sources of ions', *Int J Biometeorol*, 1963, **6**, pp 101–10.

27 Sulman, FS *et al*, 'Air ionometry of hot, dry desert winds (Sharav) and treatment with air ions of weather sensitive subjects', *Int J Biometeorol*, 1974, **18**, pp 313–18; Assael, M *et al*, 'Influence of artificial air ionization in the human electro-encephalogram', *Int J Biometeorol*, 1974, **18**, pp 306–12.

28 Sulman, FG, 1980, *op. cit.*

29 *The Effects of Weather on the Frequency and Severity of Migraine Headache in Southwestern Ontario*, Canadian Climate Report, 1980, pp 80–7.

30 Piorecky, J *et al*, 'Effect of Chinook winds on the probability of migraine headache occurrence', *Headache*, 1997, **37**, pp 153–8.

31 Rose, MS *et al*, 'The relationship between Chinook conditions and women's illness-related behaviours', *Int J Biometeorol*, 1995, **38**, pp 156–60; Verhoef, MJ *et al*, 'The relationship between Chinook conditions and women's physical and mental well being', *Int J Biometeorol*, 1995, **38**, pp 148–51.

32 Cooke, LJ *et al*, 'Chinook winds and migraine headaches', *Neurol*, 2000, **54**, pp 302–7.

33 Beasley, VR, 'Behavioral effects of air ions', *Dimension Electro Vibratory Phenomena*, 1975, **1**, pp 1–6; Sulman, FG, 1980, *op. cit.*

34 Krueger, AP, 1972. *op. cit.*

35 Fjellner, B and Hagermark, O, 'Studies on pruritogenic and histamine-releasing effects of some putative peptide neuro-transmitters', *Acta Dermatol Venereol Stockh*, 1981, **62**, pp 245–50.

36 Sutherland, ER *et al*, 'Elevated Serum Melatonin is Associated with the Nocturnal Worsening of Asthma', *J Allerg Clin Immunol*, 2003, **112**, pp 513–17.

37 Corbett, SW, 'Asthma exacerbations during Santa Ana winds in southern California', *Wilderness Environ Med*, 1996, **7**, pp 304–11.

38 Miric, D *et al*, 'Meteorologic effects and thrombocyte aggregations in patients with myocardial infarct living in the coastal region of central Dalmatia', *Lijec Vjesn*, 1993, **115**, pp 221–4.

39 Miric, D *et al*, 'The sirocco wind increases the onset of paroxysmal atrial fibrillation in patients in the central Dalmatian coastal region', *Lijec Vjesn*, 1992, **114**, pp 93–5.

40 Lewis, DL *et al*, 'Interactions of pathogens and irritant chemicals in land-applied sewage sludges (biosolids)', *BMC Public Health*, 2002, **2**, p 11.

41 Lenes, JM *et al*, Iron fertilization and the trichodesmium response on the West Florida shelf', *Limnol Oceanogr*, 2001, **46**, pp1261–1277.

42 Soyka, F and Edmonds, A, *The Ion Effect*, New York: Bantam Books, 1991.

43 Zhang, H *et al*, 'Reaction of peroxynitrite with melatonin, a mechanistic study', *Chem Res Toxicol*, 1999, **12**, pp 526–34.

44 Korszun, A *et al*, 'Melatonin levels in women with fibromyalgia and chronic fatigue syndrome', *J Rheumatol*, 1999, **26**, pp 2675–80.
45 Reiter, R, 'Atmospheric electricity and natural radioactivity', in Light, S (ed), *Medical Climatology*, Baltimore, MD: Waverly Press, 1964; see also Tromp, SW, 1980, *op. cit.*
46 Huntington, E, *Mainsprings of Civilisation*, New York: Wiley & Sons, 1945.
47 Watson, L, 1988, *op. cit.*
48 Kleinknecht, RA, 'Afraid of the weather? – Weather phobias are the often hidden consequences of severe weather phenomena', *Weatherwise*, 2002, **55**, pp 14–17.
49 Bourdon, KH *et al*, 'Gender differences in phobias, results of the ECA community survey', *J Anxiety Disord*, 1988, **2**, pp 227–41.

## Chapter 5: Stormy Weather

1 Price, SL, 'The Return', *Sports Illustrated*, 17 July 1995.
2 Affleck, G *et al*, 'Attributional processes in rheumatoid arthritis patients', *Arthritis Rhem*, 1987, **30**, pp 927–31.
3 Laborde, JM *et al*, 'Influence of weather on osteoarthritis', *Soc Sci Med*, 1986, **23**, pp 549–54.
4 Wolfe, F *et al*, 'Fibrositis: symptom frequency and criteria for diagnosis', *J Rheumatol*, 1985, **12**, pp 1159–63.
5 Katz, JL and Weiner, H, 'Psychobiological variables in the onset and recurrence of gouty arthritis, a chronic disease mode', *J Chronic Dis*, 1975, **28**, pp 51–62.
6 Kranzl, B, 'Trigeminal neuralgia, effect of external and internal influences on the disease as well as studies on infiltration treatment of pain', *Osterr Z Stomatol*, 1977, **74**, pp 246–61.
7 Krause, I *et al*, 'Seasons of the year and activity of SLE and Behcet's disease', *Scand J Rheumatol*, 1997, **26**, pp 435–39.
8 Harlfinger, O, 'Weather-induced effects on pain perception', *Fortschr Med*, 1991, **109**, pp 647–50.
9 Hoppe, P *et al*, 'Prevalence of weather sensitivity in Germany', *Deutsch Med Wochenschr*, 2002, **127**, pp 15–20.
10 Cited in Shutty, MS, 'Pain complaint and the weather, weather sensitivity and symptom complaints in chronic pain patients', *Pain*, 1992, **49**, pp 199–204.
11 Drane, D *et al*, 'The association between external weather conditions and pain and stiffness in women with rheumatoid

arthritis', *J Rheumatol*, 1997, **24**, pp 1309–16; Gorin, AA *et al*, 'Rheumatoid arthritis patients show weather sensitivity in daily life but the relationship is not significant', *Pain*, 1999, **81**, pp 173–7; van de Laar, MA *et al* 'Assessment of inflammatory joint activity in rheumatoid arthritis in atmospheric conditions', *Clin Rheumatol*, 1991, **10**, pp 426–33.

12 Hill, DF, 'Climate and arthritis in arthritis and allied conditions', in Hollander, JL and McCarthy, DC (eds) *A Textbook of Rheumatology*, 8th edn, Philadelphia: Lea and Febinger, 1972.

13 Yunus, M *et al*, 'Primary fibromyalgia, clinical study of 50 patients with matched normal controls', *Semin Arthritis Rheum*, 1981, **11**, pp 151–71.

14 Shutty, MS *et al*, 1992, *op. cit.*

15 Kasai, Y *et al*, 'Change of barometric pressure influences low back pain in patients with vacuum phenomenon within lumbar intervertebral disc', *J Spinal Disord Tech*, 2002, **15**, pp 290–3; McGorry, RW *et al*, 'Meteorological conditions and self-report of low back pain', *Spine*, 1998, **23**, pp 2096–102, p 2103; Menges, LJ, 'Chronic low back pain, a medical–psychological report', *Soc Sci Med*, 1983, **17**, pp 747–53; Hendler, NH *et al*, 'The relationship of diagnoses and weather sensitivity in chronic pain patients', *Neuromuscul Sys*, 1995, **3**, pp 10–15.

16 Anderson, B *et al*, 'Migraine-like phenomena after decompression from hyperbaric environment', *Neurology*, 1965, **15**, pp 1035–40; Nursall, A, 'The effects of weather conditions on the frequency and severity of migraine headaches in southwestern Ontario', unpublished MSc thesis, University of Alberta, 1981; see also Brown, BB, *Stress and the Art of Biofeedback*, New York: Bantam Books, 1977.

17 Raskin, NH, *Headache*, 2nd edn, New York: Churchill Livingstone, 1988.

18 Sibley, JT, 'Weather and arthritis symptoms', *J Rheumatol*, 1985, **12**, pp 707–10; De Blecourt, ACE *et al*, 'Weather conditions and complaints in fibromyalgia', *J Rheumatol*, 1993, **20**, pp 1932–34; Wilkinson, M and Woodrow, J, 'Migraine and weather', *Headache*, 1979, **19**, pp 375–8; Diamond, S *et al*, 'The effects of weather on migraine frequency in Chicago', *Headache Quarterly*, 1990, **1**, pp 136–45; Schulman, J *et al*, 'The relationship of headache occurrence to barometric pressure', *Int J Biometeorol*, 1980, **24**, pp 263–9.

19 Hawley, DJ *et al*, 'Seasonal symptoms severity in patients with rheumatic disease – a study of 1424 patients', *J Rheumatol*, 2001, **28**, pp 1900–9.

20 Rosen, S, *Weathering: How the Atmosphere Conditions Your Body, Your Mind, Your Moods – and Your Health*, New York: M. Evans & Company, 1979.

21 Redelmeier, DA and Tversky, A, 'On the belief that arthritis pain is related to the weather', *Proc Natl Acad Sci USA*, 1996, **93**, pp 2895–6.

22 Dordick, I, 'The influence of variations in atmospheric pressure upon human beings', *Weather*, 1958, **13**, pp 359–63.

23 Caplan, CE, 'The big chill, diseases exacerbated by exposure to cold', *CMAJ*, 1999, **160**, p 33; Moos, RH, *The Human Context: Environmental Determinants of Behavior*, New York: John Wiley and Sons, 1976; Garvey, L, 'The body barometer', *Health*, 1987, **19**, pp 80–5; Aikman, H, 'The association between arthritis and the weather', *Int J Biometeorol*, 1997, **40**, pp 192–9; see also Guedj, A, 1990, *op. cit.*; Hill, DF, 1972, *op. cit.*; Patberg, WR, 1985, *op. cit.*; Rasker, JJ, 1986, *op. cit.*

24 Weinbrecth, WU and Simon, F, 'Effect of meteorologic parameters on acute admission of patients with lumbar intervertebral disc displacement', *Z Orthop Ihre Grenzgeb*, 1989, **127**, pp 650–2.

25 Piorecky, J *et al*, 'Effect of chinook winds on the probability of migraine headache occurrence', *Headache*, 1997, **37**, pp 153–8; Cooke, LJ *et al* 'Chinook winds and migraine headache', *Neurology*, 2000, **54**, pp 302–7.

26 Quoted in Thomson, WAR, *A Change of Air*, London: Adam & Charles Black, 1979.

27 Guedj, D and Weinberger, A, 'Effect of weather conditions on rheumatic patients', *Ann Rheum Dis*, 1990, **49**, pp 158–9.

28 Arber, N *et al*, 'Effect of weather conditions on acute gouty arthritis', *Scand J Rheumatol*, 1994, **23**, pp 22–4.

29 Jamison, RN, 'Weather changes and pain, perceived influence of local climate on pain complaint in chronic pain patients', *Pain*, 1995, **61**, pp 309–15.

30 Patberg, WR, 'Relation between meteorological factors and pain in rheumatoid arthritis in a marine climate', *J Rheumatol*, 1985, **12**, pp 711–5.

31 Dequeker, J and Wuestenraed, L, 'The effect of biometeorological factors on Ritchie articular index and pain in rheumatoid arthritis', *Scand J Rheumatol*, 1986, **15**, pp 280–4.

32 Latman, N and Levi, LN, 'Rheumatoid arthritis and climate', *N Engl J Med*, 1980, **303**, p 1178; see also Jamison, RN *et al*, 1995, *op. cit.*

33 Patberg, WR, 'Correlation of erythrocyte sedimentation rate and outdoor temperature in a patient with rheumatoid arthritis', *J Rheumatol*, 1997, **24**, pp 1017–8.

34 Rasker, JJ *et al*, 'Influence of weather on stiffness and force in patients with rheumatoid arthritis', *Scand J Rheumatol*, 1986, **15**, pp 27–36.

35 Besson, JM and Chaouch, H, 'Peripheral and spinal mechanisms of pain', *Phys Rev*, 1987, **67**, pp 67–184; see also Rasker, JJ *et al*, 1986, *ibid*.

36 Weiner, R, 'Barodontalgia, caught between the clouds and the waves', *J Mass Dent Soc*, 2002, **51**, pp 46–9.

37 Sulman, FG, 'The impact of weather on human health', *Rev Environ Health*, 1984, **4**, pp 83–119; Romano, JM and Turner, JA, 'Chronic pain and depression, does the evidence support a relationship?', *Psychol Bull*, 1985, **97**, pp 18–34; Persinger, MA, *The Weather Matrix and Human Behavior*, New York: Praeger, 1980.

38 Landsberg, HE, *Weather and Health: An Introduction to Biometeorology*, New York: Doubleday/Anchor, 1969.

39 Sherretz, LA, 'Weather and the classroom climate', doctoral dissertation, Boulder, CO: University of Colorado, 1984.

40 Brown, GI, 'The relationship between barometric pressure and relative humidity and classroom behavior', *J Edu Res*, 1964, **57**, pp 368–70.

41 Kals, WS, *Your Health, your Moods and the Weather*, New York: Doubleday & Co, 1982; see also Landsberg, HE, 1969, *op. cit.*

42 Tromp, SW, *Medical Biometeorology*, Amsterdam: Elsevier Publishing Company, 1963.

43 Tromp, SW, *Biometeorology: The Impact of the Weather and Climate on Humans and their Environment*, London: Heyden & Son Ltd, 1980.

44 Harlfinger, O, 'Weather sensitivity among elementary school children', in *Proceedings of the 13th International Congress of Biometeorology*, 12–18 September 1993, Calgary, Alberta, Canada.

45 Delyukov, A and Didyk, L, 'The effects of extra-low-frequency atmospheric pressure oscillations on human mental activity', *Int J Biometeorol*, 1999, **43**, pp 31–37.

46 Sulman, FG *et al*, 'Urinalysis of patients suffering from climatic heat stress', *Int J Biometeorol*, 1970, **14**, pp 45–53.

47 Rosenthal, N, *Seasons of the Mind: Why You Get the Winter Blues*, New York: Bantam Books, 1990.

48 Marks, GB *et al*, 'Thunderstorm outflows preceding epidemics of asthma during spring and summer', *Thorax*, 2001, **56**, pp 468–71.

49 Reiter, R, *Phenomena In Atmospheric And Environmental Electricity*, Amsterdam: Elsevier, 1992.

50 Hoffmann, G *et al*, 'Significant correlations between certain spectra of atmospherics and different biological and patholoical parameters', *Int J Biometeorol*, 1991, **34**, pp 247–50; Eichmeier, J and Baumer, H, *Atmospherics Emission Computer Tomography and its Importance for Biometeorology*, Abstracts, 11th Congress of Biometeorology, Purdue University, West Laffayette, USA, The Hague: SPB Academic Publishing, 1987.

51 Reiter, R, *Meteorobiology And Atmospheric Electricity*, Leipzig: Akademische Verlagsgesellschaft Geest & Portig, 1960.

52 Schienle, A *et al*, 'Biological effects of very low frequency (VLF) atmospherics in humans: a review', *J Sci Explor*, 1998, **12**, pp 455–68.

53 Newson, R *et al*, 'Effect of thunderstorms and airborne grass pollen on the incidence of acute asthma in England, 1990–94', *Thorax*, 1997, **52**, pp 680–85.

54 Newson, R *et al*, 'Acute asthma epidemics, weather and pollen in England, 1987–1994', *Eur Respir J*, 1998, **11**, pp 694–701.

55 Buettner, KJK, 'Present knowledge on correlations between weather changes, sferics, and air electric space charges and human health and behavior', *Fed Proc*, 1957, **16**, pp 631–37; Friedman, H and Becker, RO, 'Geomagnetic parameters and psychiatric hospital admissions', *Nature*, 1963, **200**, pp 626–8; Krueger, AP and Smith, RF, 'An enzymatic basis for the acceleration of ciliary activity by negative air ions', *Nature*, 1959, **183**, pp 1332–33; Schaefer, HJ, 'Man and radiant energy – ionizing radiation', in *Handbook Of Physiology*, 'Section 4: Adaptation to the environment', Washington DC: American Physiological Society, 1964.

56 Ruhenstroth-Bauer, G *et al*, 'Myocardial infarction and the weather: a significant positive correlation between the onset of heart infarct and 28 KHz atmospherics – a pilot study', *Clin Cardiol*, 1985, **8**, pp 149–51.

57 Jacobi, E *et al*, 'Simulated VLF-fields as a risk factor of thrombosis', *Int J Biometeorol*, 1981, **25**, pp 133–42; Sulman, FG, *The Effect of Air Ionisation, Electrical Fields, Atmospherics and Other Electric Phenomena on Man and Animal*, Springfield, IL: Charles C Thomas, 1980.

58 Schienle, A *et al*, 'Electrocortical responses of headache patients to the simulation of 1-kHz sferics', *Int J Neurosci*, 1999, **97**,

pp 211–24; Vaitl, D *et al*, 'Headache and sferics', *Headache*, 2001, **41**, pp 845–53.

59 Delyukov, AA *et al*, 'Natural environmental associations in a 50-day human electrocardiogram', *Int J Biometeorol*, 2001, **45**, pp 90–99.

60 Stump, B, 'Under the weather?' *Men's Health*, 1999, **14**, pp 124–41.

61 Matsushima, S *et al*, 'Effect of magnetic field on pineal gland volume and pinealocycte size in the rat', *J Pineal Res*, 1993, **14**, pp 145–50; Lerchl, A *et al*, 'Marked rapid alterations in noctural pineal serotonin metabolism in mice and rats exposed to weak intermittent magnetic fields', *Biochem Biophys Res Commun*, 1990, **169**, pp 102–8.

62 Pilla, AA and Markov, MS, 'Bioeffects of weak electromagnetic fields', *Rev Environ Health*, 1994, **10**, pp 155–69.

63 Schienle, A *et al*, 'Effects of low frequency magnetic fields on electrocortical activity in humans: A sferics simulation study', *Int J Neurosci*, 1997, **90**, pp 21–36.

64 Reiter, R, *Phenomena in Atmospheric and Environmental Electricity*, Amsterdam: Elsevier, 1992.

65 Schienle, A *et al*, 'Atmospheric electromagnetism: Individual difference into brain electrical response to stimulated sferics', *Int J Psychophysiol*, 1996, **21**, pp 177–88.

66 Gossard, EE and Hooke, WH, *Waves in the Atmosphere, Atmospheric Infrasound and Gravity Waves – Their Generation and Propagation*, Amsterdam, Oxford, New York: Elsevier Scientific, 1975.

67 Broner, N, 'The effects of low frequency noise on people – a review', *J Sound Vibr*, 1978, **58**, pp 483–500; Bull, G *et al*, 'Infrasonic waves in the atmosphere', *Z Meteorol*, 1988, **38**, pp 265–83.

68 Karpova, NI *et al*, 'Early response of the organism to low-frequency acoustic oscillations', *Noise Vib Bull*, 1970, **11**, pp 100–3; Kawano, A *et al*, 'Effects of infrasound on humans: A questionnaire survey of 145 drivers of long distance transport trucks', *Pract Otol Kyoto*, 1991, **84**, pp 1315–24; Delyukov, A, 1999, *op. cit.*

69 Hecht, J, 'Not a sound idea', *New Scientist*, 20 March 1999. Available at http://trauma.cofa.unsw.edu.au/Infrasound/NewScientist01.html.

70 Swanson, DC, 'Non-lethal acoustic weapons: Facts, fiction, and the future', abstract of presentation at the NTAR 1999 Symposium. Available at http://www.unh.edu/orps/nonlethality/pub/abstracts/1999/swanson.html.

71 Altmann, J, 'Acoustic weapons', 1999. Available at http://trauma.cofa.unsw.edu.au/Infrasound/acousticweapons.pdf.

72 Broner, N, 'The effects of low frequency noise on people – a review', *J Sound Vibr*, 1978, **58**, pp 483–500; Andresen, J and Moller, H, 'Equal annoyance contours for infrasonic frequencies', *J Low Freq Noise Vibr*, 1984, **3**, pp 1–9.
73 Arabadzhi, VI, 'Infrasound and biorhythms of human brain', *Biophyzika*, 1992, **37**, pp 150–1.
74 Landstrom, U and Bystrom, M, 'Infrasonic threshold levels of physiological effects', *J Low Freq Noise Vibr*, 1984, **3**, pp 167–73.
75 Green, JE and Dunn, F, 'Correlation of naturally occurring infrasonics and selected human behaviour', *J Acoust Soc Am*, 1968, **44**, pp 1456–57.
76 Kompanets, VN, 'The effect on human of various changes in barometric pressure', *Voen Med Zhurn*, 1968, **6**, pp 61–63.
77 Gavreau, V, 'Infrasound', *Science J*, 1968, **4**, pp 33–37.
78 Richner, H and Graber, W, 'The ability of non-classical meteorological parameters to penetrate into buildings', *Int J Biometeorol*, 1978, **22** (2), pp 242–48.

## Chapter 6: Having a Heatwave

1 Utiger, RD, 'The need for more vitamin D', *N Engl J Med*, 1998, **338**, pp 828–9.
2 Grimes, D *et al*, 'Sunlight, cholesterol and coronary heart disease', *QJ Med*, 1996, **89**, pp 579–89.
3 Guyton, KZ *et al*, 'Vitamin D and vitamin D analogs as cancer chemopreventive agents', *Nutr Rev*, 2003, **61**, pp 227–38.
4 Arthey, S and Clarke, VA, 'Suntanning and sun protection: A review of the psychological literature', *Soc Sci Med*, 1995, **40**, pp 265–74.
5 Anderson, CA, 'Temperature and aggression: Ubiquitous effects of heat on occurrence of human violence', *Psychol Bull*, 1989, **106**, pp 74–96.
6 Reifman, AS *et al*, 'Temper and temperature on the diamond: The heat–aggression relationship in major league baseball', *Pers Soc Psychol Bull*, 1991, **17**, pp 580–85.
7 Baron, RA and Bell, PA, 'Aggression and heat: The influence of ambient temperature, negative affect, and a cooling drink on physical aggression', *J Pers Soc Psychol*, 1976, **33**, pp 245–55.
8 Kenrick, DT and MacFarlane, SW, 'Ambient temperature and horn-honking: A field study of the heat/aggression relationship', *Environ Behav*, 1984, **18**, pp 179–91.

9 Robbins, MC *et al*, 'Climate and behavior: A biocultural study', *J Cross-Cultural Psychol*, 1972, **3**, pp 331–44.
10 Cotton, JL, 'Ambient temperature and violent crime', *J Appl Soc Psychol*, 1986, **16**, pp 786–801; Tennenbaum, AN and Fink, EL, 'Temporal regularities in homicide, cycles, seasons and autoregression', *J Quant Criminol*, 1994, **10**, pp 317–42.
11 Michael, RP and Zumpe, D, 'Sexual violence in the United States and the role of season', *Am J Psychiatr*, 1983, **140**, pp 883–6; Anderson, CA, 'Temperature and aggression: Effects on quarterly, yearly, and city rates of violent and nonviolent crime', *J Pers Soc Psychol*ogy, 1987, **52**, pp 1161–73: Schreiber, G *et al*, 'Photoperiodicity and annual rhythms of wars and violent crimes', *Med Hypoth*, 1997, **48**, pp 89–96.
12 Rotton, J and Frey, J, 'Air pollution, weather, and violent crimes: Concomitant time-series analysis of archival data', *J Pers Soc Psychol*, 1985, **49**, pp 1207–20.
13 Michael, RP and Zumpe, D, 'An annual rhythm in the battering of women', *Am J Psychiatry*, 1986, **143**, pp 637–40.
14 Michael, RP and Zumpe, D, 'Sexual violence in the United States and the role of season', *Am J Psychiatry*, 1983, **140**, pp 883–6.
15 Anderson, CA, 'Temperature and aggression: Effects on quarterly, yearly, and city rates of violent and nonviolent crime', *J Pers Soc Psychol*, 1987, **52**, pp 1161–73.
16 Anderson, CA and Anderson, DC, 'Ambient temperature and violent crime: Tests of the linear and curvilinear hypothesis', *J Pers Soc Psychol*ogy, 1984, **46**, pp 91–7.
17 Durkheim, E, *Suicide: A Study In Sociology* (translated from Durkeim, E, *Le Suicide*, Paris, 1897), London: Lowe and Brydone Ltd, 1970.
18 Kevan, SM, 'Perspectives on season of suicide: A review', *Soc Sci Med*, 1980, **14**, pp 369–378; Massing, W and Angermeyer, MC, 'The monthly and weekly distribution of suicide', *Soc Sci Med*, 1985, **21**, pp 433–41; Chew, KSY and McCleary, R, 'The spring peak in suicides: A cross-national analysis', *Soc Sci Med*, 1995, **40**, pp 223–230; Altamura, C *et al*, 'Seasonal and circadian rhythms in suicide in Calgari, Italy', *J Affect Disord*, 1999, **53**, pp 77–85.
19 Parker, G and Walter, S, 'Seasonal variation in depressive disorders and suicidal deaths in New South Wales', *Br J Psychiatry*, 1982, **40**, pp 626–32; Flisher, AJ *et al*, 'Seasonal variation of suicide in South Africa', *Psychiatry Res*, 1997, **66**, pp 13–22.

20 Souetre, E *et al*, 'Seasonality of suicides, environmental, sociological and biological covariations', *J Affect Disord*, 1987, **13**, pp 215–25; Salib E and Gray, N, 'Weather conditions and fatal self-harm in North Cheshire 1989–1993', *Br J Psychiatry*, 1997, **170**, pp 473–7; Preti, A, 'The influence of seasonal change on suicidal behaviour in Italy', *J Affect Disord*, 1997, **44**, pp 123–30.

21 Näyhä, S, 'Autumn incidence of suicides re-examined: Data from Finland by sex, age and occupation', *Br J Psychiatry*, 1982, **141**, pp 512–7; Eastwood, MR and Peacocke, J, 'Seasonal patterns of suicide, depression and electroconvulsive therapy', *Br J Psychiatry*, 1976, **129**, pp 472–5; Maes, M *et al*, 'Seasonality in severity of depression, relationships to suicide and homicide occurrence', *Acta Psychiatr Scand*, 1993, **88**, pp 156–61; Parker, G and Walter, S, 1982 *et al*, *op. cit.*

22 Jessen, G *et al*, 'Attempted suicide and major public holidays in Europe: Findings from the WHO/EURO multicentre study on parasuicide', *Acta Psychiatr Scand*, 1999, **99**, pp 412–8; Nakamura, JW *et al*, 'Temporal variation in adolescent suicide attempts', *Suicide Life Threat Behav*, 1994, **24**, pp 343–9.

23 Maes, M *et al*, 'Biochemical, metabolic and immune correlates of seasonal variation in violent suicide: A chronoepidemiologic study', *Eur J Psychiat*, 1996, **11**, pp 21–33; Maes, M *et al*, 'Seasonal variation in plasma L-tryptophan availability in healthy volunteers, relationships to violent suicide occurrence', *Arch Gen Psychiatry*, 1995, **52**, pp 937–46.

24 Fossey, E and Shapiro, CM, 'Seasonality in psychiatry – a review', *Can J Psychiatry*, 1992, **37**, pp 299–308; Maes, M *et al*, 'Synchronized annual rhythms in violent suicide rate, ambient temperature and the light–dark span', *Acta Psychiatr Scand*, 1994, **90**, pp 391–6; Altamura, C *et al*, 1999, *op. cit.*

25 Parsons, AG, 'The association between daily weather and daily shopping patterns', *Australasian Marketing J*, 2001, **9**, pp 78–84.

26 Rind, B, 'Effects of beliefs about weather conditions on tipping', *J Appl Soc Psychol*, 1996, **26**, pp 137–47.

27 Keim, DB, 'Size-related anomalies and stock return seasonality: Further evidence', *J Financial Econ*, 1983, **12**, pp 13–32.

28 Saunders, EM, 'Stock prices and Wall Street weather', *Am Econ Rev*, 1993, **83**, pp 1337–45.

29 Quoted in Cuvelier, M, 'Does the sun skew your judgment? A hidden influence on the stock market', *Psychol Today*, 2001, **34**, p 24.

30 Bazett, HC, 'Physiological responses to heat', *Physiological Rev*, 1927, **7**, pp 531–99; Hardy, JD, 'Physiology of temperature regulation', *Physiological Rev*, 1961, **41**, pp 521–605; Tromp, SW, *Biometeorology: The Impact of the Weather and Climate on Humans and Their Environment*, London: Heydon, 1980; Bloch, G, *Body and Self: Elements of Human Biology – Behavior and Health*, Los Altos, CA: Kaufman, 1985; Oken, D *et al*, 'Relation of physiological responses to affect expression', *Arch Gen Psychiatry*, 1962, **6**, pp 336–52; Persinger, MA, *The Weather Matrix and Human Behavior*, New York: Praeger, 1980.

31 Zajonc, RB, 'Emotion and facial efference: A theory re-examined', *Science*, 1985, **228**, pp 15–21; Zajonc, RB, 'Emotional expression and temperature modulation', in van Goozen, SHM *et al* (eds), *Emotions: Essays On Emotional Theory*, Hillsdale, NJ: Lawrence Erlbaum Assoc, 1984; Zajonc, RB *et al*, 'Feeling and facial efference: Implications of the vascular theory of emotion', *Psychological Rev*, 1989, **96**, pp 395–416.

32 *Ibid.*

33 Olweus, D *et al*, 'Testosterone, aggression, physical, and personality dimensions in normal adolescent males', *Psychosom Med*, 1980, **42**, pp 253–69; Palamarek, DL and Rule, BG, 'The effects of ambient temperature and insult on the motivation to retaliate or escape', *Motiv Emotion*, 1979, **3**, pp 83–92.

34 Andersson, A *et al*, 'Variation in levels of serum inhibin B, testosterone, estradiol, luteinizing hormone, follicle-stimulating hormone, and sex hormone-binding globulin in monthly samples from healthy men during a 17–month period: Possible effects of seasons', *J Clin Endocrinol Metab*, 2003, **88**, pp 932–937.

35 Virkkunen, M *et al*, 'CSF biochemistries, glucose metabolism, and diurnal activity rhythms in alcoholic, violent offenders, firesetters, and healthy volunteers', *Arch Gen Psychiatry*, 1994, **51**, pp 20–27.

36 Bligh, J, *Temperature Regulation in Mammals and Other Vertebrates*, New York: American Elsevier, 1973.

37 Lambert, G *et al*, 'Increased suicide rate in the middle-aged and its association with hours of sunlight', *Am J Psychiatry*, 2003, **160**, pp 793–5; Lambert, GW *et al*, 'Effect of sunlight and season on serotonin turnover in the brain', *Lancet*, 2002, **360**, pp 1840–2.

38 Lidberg, L *et al*, 'Homicide, suicide and CSF 5–HIAA', *Acta Psychiatr Scand*, 1985, **71**, pp 230–6.

39 Brown, GL *et al*, 'Aggression in humans correlates with cerebrospinal fluid amine metabolites', *Psychiatry Res*, 1979, **1**,

pp 131–9; Linnoila, M *et al*, 'Low cerebrospinal fluid 5-hydroxyin doleaceti acid concentration differentiates impulsive from nonimpulsive violent behaviour', *Life Sci*, 1983, **33**, pp 2609–14; Linnoila, VM and Virkkunen, M, 'Aggression, suicidality and serotonin', *J Clin Psychiatry*, 1992, **53**, pp 46–51.

40 D'Mello, DA *et al*, 'Seasons and bipolar disorder', *Ann Clin Psychiatry*, 1995, **7**, pp 11–8; Faedda, GL *et al*, 'Seasonal mood disorders: Patterns of seasonal recurrence in mania and depression', *Arch Gen Psychiatry*, 1993, **50**, pp 17–23.

41 Fornari, VM *et al*, 'Seasonal patterns in eating disorder subgroups', *Comp Psychiatry*, 1994, **35**, pp 450–56; Levitan, RD *et al*, 'Seasonal fluctuations in mood and eating behaviour in bulimia nervosa', *Int J Eat Disord*, 1994, **16**, pp 295–99.

42 Lam, RW *et al* 'Seasonal symptoms in anorexia and bulimia nervosa', *Int J Eat Disord*, 1996, **19**, pp 35–44.

43 Sloan, DM, 'Does warm weather climate affect eating disorder pathology?', *Int J Eat Disord*, 2002, **32**, pp 240–44.

44 Martin, N, 'Summer seasonal affective disorder', *Nurs Stand*, 1992, **6**, pp 32–5.

45 Wehr, TA and Rosenthal, NE, 'Seasonality and affective illness', *Am J Psychiatry*, 1989, **146**, pp 829–39.

46 Goyer, PF *et al*, 'Cerebral glucose metabolism in patients with summer seasonal affective disorder', *Neuropsychopharmacol*, 1992, **7**, pp 233–40.

47 Wehr, TA *et al*, 'Summer depression: Description of the syndrome and comparison with winter depression', in Rosenthal, NE and Blehar, MC (eds), *Seasonal Affective Disorders and Phototherapy*, New York: Guilford Press, 1989.

48 Scotto, J and Nam, J, 'Skin melanoma and seasonal patterns', *Am J Epidemiol*, 1980, **118**, pp 785–6: Cohen, P, 'Cancer and seasonal patterns', *Am J Epidemiol*, 1983, **118**, pp 785–6; Cohen, P *et al*, 'Seasonality in the occurrence of breast cancer', *Cancer Res*, 1983, **43**, pp 892–96.

49 Yoshikawa, T *et al*, 'Susceptibility to effects of UVB radiation on induction of contact hypersensitivity as a risk factor for skin cancer in humans', *J Invest Dermatol*, 1990, **95**, pp 530–6.

50 De Hertog, SA *et al*, 'Relation between smoking and skin cancer', *J Clin Oncol*, 2001, **19**, pp 231–8.

51 Kearney, R, 'Promotion and prevention of tumour growth – effects of endotoxin, inflammation and dietary lipids', *Int Clin Nutr Rev*, 1987, **7**, p 157.

52 Wolk, A *et al*, 'A prospective study of association of monounsaturated fat and other types of fat with risk of breast cancer', *Arch Intern Med*, 1998, **158**, pp 41–5.
53 Mackie, BS, 'Malignant melanoma and diet', *Med J Austr*, 1974, **1**, p 810.
54 Mackie, BS *et al*, 'Melanoma and dietary lipids', *Nutr Cancer*, 1987, **9**, pp 219–26.
55 Veierod, MB *et al*, 'Diet and risk of cutaneous malignant melanoma: A prospective study of 50,757 Norwegian men and women', *Int J Cancer*, 1997, **71**, pp 600–4.
56 Adam, J, 'Sun-protective clothing', *J Cutan Med Surg*, 1998, **3**, pp 50–3.
57 Gies, HP *et al*, 'Textiles and sun protection', in Volkmer, B *et al* (eds), *Environmental UV Radiation, Risk of Skin Cancer and Primary Prevention*, Stuttgart: Gustav Fischer, 1996.
58 Gambichler, T *et al*, 'Protection against ultraviolet radiation by commercial summer clothing: Need for standardised testing and labelling', *BMC Dermatol*, 2001, **1**, p 6.
59 Garland, CF *et al*, 'Could sunscreens increase melanoma risk?', *Am J Public Health*, 1992, **82**, pp 614–5.
60 Green, A *et al*,'Daily sunscreen application and betacarotene supplementation in prevention of basal-cell and squamous-cell carcinomas of the skin: A randomised controlled trial', *Lancet*, 1999, **354**, pp 723–9; Dover, JS and Arndt, KA, 'Dermatology', *JAMA*, 1994, **271**, pp 1662–3.
61 Garland, CF *et al*, 'Could sunscreens increase melanoma risk?', *Am J Pub Health*, 1992, **82**, pp 614–15; Garland, CF *et al*, 'Effect of sunscreens on UV radiation-induced enhancement of melanoma growth in mice', *J Natl Cancer Inst*, 1994, **86**, pp 798–801.
62 Vainio, H *et al*, 'An international evaluation of the cancer-preventive potential of sunscreens', *Int J Cancer*, 2000, **88**, pp 838–42; Garland, CF *et al*, 'Rising trends in melanoma. A hypothesis concerning sunscreen effectiveness', *Ann Epidemiol*, 1993, **3**, pp 103–10; Weinstock, MA, 'Do sunscreens increase or decrease melanoma risk: An epidemiologic evaluation', *J Investig Dermatol Symp Proc*, 1999, **4**, pp 97–100: Fleming, C and Mackie, R, 'Sunscreens, suntans and skin cancer. Knowledge about sunscreens is inadequate', *BMJ*, 1996, **313**, p 942: La Vecchia, C, 'Sunscreens and the risk of cutaneous malignant melanoma', *Eur J Cancer Prev*, 1999, **8**, pp 267–9.
63 Bastuji-Garin, S and Diepgen, TL, 'Cutaneous malignant melanoma, sun exposure, and sunscreen use: Epidemiological

evidence', *Br J Dermatol*, 2002, **146**, pp 24–30; Huncharek, M and Kupelnick, B, 'Use of topical sunscreen and the risk of malignant melanoma. Results of a meta–analysis of 9,067 patients', *Ann Epidemiol*, 2000, **10**, p 467.

64 Emmons, KM and Colditz, GA, 'Preventing excess sun exposure: It is time for a national policy', *J Natl Cancer Inst*, 1999, **91**, pp 1269–70.

65 Gulston, M and Knowland, J, 'Illumination of human keratinocytes in the presence of the sunscreen ingredient Padimate-O and through an SPF-15 sunscreen reduces direct photodamage to DNA but increases strand breaks', *Mutat Res*, 1999, **444**, pp 49–60; Cockell, CS and Knowland, J, 'Ultraviolet radiation screening compounds', *Biol Rev Camb Philos Soc*, 1999, **74**, pp 311–45; Dunford, R *et al*, 'Chemical oxidation and DNA damage catalysed by inorganic sunscreen ingredients', *FEBS Lett*, 1997, **418**, pp 87–90; Knowland, J, 'Sunlight-induced mutagenicity of a common sunscreen ingredient', *FEBS Lett*, 1993, **324**, pp 309–13; Schallreuter, KU *et al*, 'Oxybenzone oxidation following solar irradiation of skin: Photoprotection versus antioxidant inactivation', *J Invest Dermatol*, 1996, **106**, pp 583–6.

66 Pasco, JA *et al*, 'Vitamin D status of women in the Geelong Osteoporosis Study: Association with diet and casual exposure to sunlight', *Med J Aust*, 2001, **175**, pp 401–5; McGrath, JJ *et al*, 'Vitamin D insufficiency in south-east Queensland', *Med J Aust*, 2001, **174**, pp 150–1.

67 Mason, RS and Diamond, TH, 'Vitamin D deficiency and multicultural Australia', *Med J Aust*, 2001, **175**, pp 236–7.

68 Garland, FC *et al*, 'Geographic variation in breast cancer mortality in the United States: A hypothesis involving exposure to solar radiation', *Prev Med*, 1990, **19**, pp 614–22.

69 Lucas, RM *et al*, *Comparative Risk Assessment: Ultraviolet Radiation*, World Health Organization, 2003.

70 McGrath, J, 'Hypothesis: Is low prenatal vitamin D a risk-modifying factor for schizophrenia?', *Schizophr Res*, 1999, **40**, pp 173–77.

71 Studzinski, GP and Moore, DC, 'Sunlight – can it prevent as well as cause cancer?', *Cancer Res*, 1995, **55**, pp 4014–22.

72 Ponsonby, AL *et al*, 'Ultraviolet radiation and autoimmune disease: Insights from epidemiological research', *Toxicology*, 2002, **181–182**, pp 71–8; Norris, JM, 'Can the sunshine vitamin shed light on type 1 diabetes?', *Lancet*, 2001, **358**, pp 1476–8.

73 Hypponen, E *et al*, 'Intake of vitamin D and risk of type 1 diabetes: A birth-cohort study', *Lancet*, 2001, **358**, pp 1500–3; Eurodiab substudy 2 study group, 'Vitamin D supplement in early childhood and risk for type 1 (insulin-dependent) diabetes mellitus', *Diabetologia*, 1999, **42**, pp 51–4.

74 Hammond, SR *et al*, 'The age-range of risk of developing multiple sclerosis. Evidence from a migrant population in Australia', *Brain*, 2000, **123**, pp 968–74; Gale, CR and Martyn, CN, 'Migrant studies in multiple sclerosis', *Prog Neurobiol*, 1995, **47**, pp 425–48; Visscher, BR *et al*, 'Latitude, migration and the prevelance of multiple sclerosis', *Am J Epidemiol*, 1977, **106**, pp 470–5.

75 van der Mei, IA *et al*, 'Regional variation in multiple sclerosis prevalence in Australia and its association with ambient ultraviolet radiation', *Neuroepidemiol*, 2001, **20**, pp 168–74.

76 McMichael, AJ and Hall, AJ, 'Does immunosuppressive ultraviolet radiation explain the latitude gradient for multiple sclerosis?', *Epidemiol*, 1997, **8**, pp 642–5.

77 Grant, WB, 'An estimate of premature cancer mortality in the US due to inadequate doses of solar ultraviolet-B radiation', *Cancer*, 2002, **94**, pp 1867–75.

78 Koh, HK and Lew, RA, 'Sunscreens and melanoma: Implications for prevention', *J Natl Cancer Inst*, 1994, **86**, pp 78–9.

79 Ainsleigh, HG, 'Beneficial effects of sun exposure on cancer mortality', *Prev Med*, 1993, **22**, pp 132–40.

80 Grant, WB, 2002, *op. cit.*

81 Ainsleigh, H, 1993, *op. cit.*

82 Holick, MF, 'Sunlight "Dilemma": Risk of skin cancer or bone disease and muscle weakness', *Lancet*, 2001, **357**, pp 4–6.

83 Kricker, A *et al*, 'A dose-response curve for sun exposure and basal cell carcinoma', *Int J Cancer*, 1995, **60**, pp 482–488.

84 Bosch, X, 'European heatwave causes misery and deaths', *Lancet*, 2003, **362**, p 543.

85 Anonymous, 'Heat-related illness and deaths – United States, 1994–1995', *Morbid Mortal Wkly Rep*, 1995, **44**, pp 465–8.

86 Kunst, AE *et al*, 'Outdoor air temperature and mortality in the Netherlands: A time-series analysis', *Am J Epidemiol*, 1993, **137**, pp 331–41.

87 Folk, GE, *Textbook of Environmental Physiology*, Philadelphia, PA: Lee and Febiger, 1974; Wyndham, CH *et al*, 'Tolerance times of high wet bulb temperatures in acclimatized and unacclimatized men', *Environment Res*, 1970, **3**, pp 339–52.

88 Rooney, C *et al*, 'Excess mortality in England and Wales, and in Greater London, during the 1995 heatwave', *J Epidemiol Commun Health*, 1998, **52**, pp 482–6.

89 Lee-Chiong, TL and Stitt, JT, 'Heatstroke and other heat-related illnesses: The maladies of summer', *Postgrad Med*, 1995, **98**, pp 26–36; Sandor, RP, 'Heat illness', *Physician Sports Med*, 1997, **25**, pp 35–40; Simon, HB, 'Hyperthermia', *N Engl J Med*, 1993, **329**, pp 483–7; Hassanein, T *et al*, 'Heatstroke: Its clinical and pathological presentation, with particular attention to the liver', *Am J Gastroenterol*, 1992, **87**, pp 1382–9; Bross, MH *et al*, 'Heat emergencies', *Am Fam Physician*, 1994, **50**, pp 389–96; Delaney, KA, 'Heatstroke: Underlying processes and lifesaving management', *Postgrad Med*, 1992, **91**, pp 379–88; Dixit, S *et al*, 'Epidemic heatstroke in a midwest community: Risk factors, neurological complications and sequelae', *Wis Med J*, 1997, **96**, pp 39–41; Mellion, MB and Shelton, GL, 'Safe exercise in the heat and heat injuries', in Mellion, MB *et al* (eds), *The Team Physician's Handbook*, 2nd ed, Philadelphia: Hanley and Belfus, 1997; Squire, DL, 'Heat illness. Fluid and electrolyte issues for pediatric and adolescent athletes', *Pediatr Clin North Am*, 1990, **37**, pp 1085–109; Tek, D and Olshaker, JS, 'Heat illness', *Emerg Med Clin North Am*, 1992, **10**, pp 299–310.

90 Schuman, SH *et al*, 'Epidemiology of successive heat waves in Michigan in 1962 and 1963', *JAMA*, 1964, **189**, pp 733–8; Jones, T *et al*, 'Morbidity and mortality associated with the July 1980 heatwave in St Louis, and Kansas City, Missouri', *JAMA*, 1982, **247**, pp 3327–31.

91 Crowe, JP and Moore, RE, 'Physiological and behavioral responses of aged men to passive heating', *J Physiol*, 1973, **236**, p 43; see also Ellis and Nelson, 1978, *op. cit.*

92 Sartor, F *et al*,'Temperature, ambient ozone levels, and mortality during summer 1994 in Belgium', *Environment Res*, 1995, **70**, pp 105–113; Semenza, JC *et al*, 'Heat-related deaths during the July 1995 heatwave in Chicago', *N Engl J Med*, 1996, **335**, pp 84–90; Kilbourne, EM, 'Heatwaves' in Noji, E (ed), *The Public Health Consequences of Disasters*, Oxford: Oxford University Press, 1997.

93 Bridger, CA *et al*, 'Mortality in St Louis, Missouri during heat waves in 1936, 1953, 1954, 1955 and 1966', *Environ Res*, 1976, **12**, pp 38–48; Oechsli, FW and Buechley, RW, 'Excess mortality

associated with three Los Angeles September hot spells', *Environ Res*, 1970, **3**, pp 277–84; Ellis, FP and Nelson, F, 'Mortality in the elderly in a heatwave in New York City, August 1975', *Environ Res*, 1978, **15**, pp 504–12; Ellis, FP *et al*, 'Mortality during heatwaves in New York City, July 1972 and August and September 1973', *Environ Res*, 1975, **10**, pp 1–13.
94 Henschel, A *et al*, 'An analysis of the heat deaths in St Louis during July 1966', *Am J Pub Health*, 1969, **59**, pp 2232–42.
95 Gover, M, 'Mortality during periods of excessive temperature', *US Public Health Reports*, 1936, **53**, pp 1112–43; Schuman, SH *et al*, 'Epidemiology of successive heatwaves in Michigan in 1962 and 1963', *JAMA*, 1964, **189**, pp 733–8.
96 Schuman, SH, 'Patterns of urban heatwave deaths and implications for prevention: Data from New York and St. Louis during July 1966', *Environ Res*, 1972, **5**, pp 59–75; Clarke, JF, 'Some climatological aspects of heat waves in the contiguous United States', *Environ Res*, 1972, **5**, pp 76–84; Henschel, A *et al*, 1969, *op. cit.*
97 Kilbourne, EM *et al*, 'Risk factors for heatstroke – a case-control study', *JAMA*, 1982, **247**, pp 3332–36.
98 Buechley *et al*, 1972, *op. cit.*
99 Marmor, M, 'Heatwave mortality in New York City 1949–1970', *Arch Environ Health*, 1975, **30**, pp 130–6.

## Chapter 7: Winter Chills

1 Gribbin, J and Gribbin, M, *Ice Age – How a Change of Climate Made Us Human*, London: Penguin, 2001.
2 de Castro, JM, 'Seasonal rhythms of human nutrient intake and meal pattern', *Physiol Behav*, 1991, **50**, pp 243–8.
3 National Center for Health Statistics, 'Standardized micro-data tape transcripts', US Department of Health, Education and Welfare, 1978, DHEW Publication No. (PHS), pp 78–1213.
4 US Department of Commerce (NOAA), *Report of the Increase in Senior Citizen Fatalities Attributable to the Severe Cold During the Last Half of December 1983*, Unpublished, 1984.
5 Kalkstein, LS, 'The impact of winter weather on human mortality', in *Climate Impact Assessment*, United States Department of Commerce, December 1984.
6 Kalkstein, LS and Davis, RE, 'The development of a weather/mortality model for environmental impact assessment',

*Proceedings of the 7th Conference of Biometeorology and Aerobiology*, 1985, pp 334–36.

7 Radomski, MW and Boutelier, C, 'Hormone response of normal and intermittent cold preadapted humans to continuous cold', *J Appl Physiol*, 1982, **53**, pp 610–16.

8 Keatinge, WR *et al*, 'Changes in seasonal mortalities with improvement in home heating in England and Wales from 1964 to 1984', Int J Biometeorol, 1989, **33**, pp 71–6.

9 Marchant, B *et al*, 'Circadian and seasonal factors in the pathogenesis of acute myocardial infarction – the influence of environmental temperature', *BMJ*, 1993, **69**, pp 385–7; Anderson, TW and le Riche, WH, 'Cold weather and myocardial infarction', *Lancet*, 1970, **1**, pp 291–6; McKee, CM, 'Deaths in winter in Northern Ireland: the role of low temperature', *Ulster Med J*, 1990, **59**, pp 17–22.

10 Douglas, AS *et al*, 'Seasonal, regional and secular variations of cardiovascular and cerebrovascular mortality in New Zealand', *Aust New Zeal J Med*, 1990, **20**, pp 669–76.

11 Scragg, R, 'Seasonality of cardiovascular disease mortality and the possible protective effect of ultra-violet radiation', *Int J Epidemiol*, 1981, **10**, pp 337–41; Behar, S, 'Out-of-hospital death in Israel – should we blame the weather?', *Isr Med Assoc J*, 2002, pp 56–57; Caplan, CE, 'The big chill: diseases exacerbated by exposure to cold', *Can Med Assoc J*, 1999, **160**, pp 88–9; Blindauer, KM *et al*, 'The 1996 New York blizzard: impact on non-injury emergency visits', *Am J Emerg Med*, 1999, **17**, pp 23–7.

12 Hajat, S and Haines, A, 'Associations of cold temperatures with GP consultations for respiratory and cardiovascular disease amongst the elderly in London', *Int J Epidemiol*, 2002, **31**, pp 825–830.

13 Leviton, R, 'How the weather affects your health', *East West*, 1989, **19**, pp 64–71; Smith, R, 'Doctors and climate change: action is needed because of the high probability of serious harm to health', *BMJ*, 1994, **309**, pp 1384–6.

14 Smith, R, 1994, *op. cit.*; Blindauer, KM, 1996, *op. cit.*

15 Douglas, AS *et al*, 'Composition of seasonality of disease', *Scott Med J*, 1991, **36**, pp 76–82; Shinkawa, A *et al*, 'Seasonal variation in stroke incidence in Hisayama, Japan', *Stroke*, 1990, **21**, pp 1262–7.

16 McKee, CM, 'Deaths in winter in Northern Ireland: the role of low temperature', *Ulster Med J*, 1990, **59**, pp 17–22;

Marchant, B *et al*, 'Circadian and seasonal factors in the pathogenesis of acute myocardial infarction – the influence of environmental temperature', *BMJ*, 1993, **69**, pp 385–387.

17 Capon, A *et al*, 'Seasonal variation of cerebral haemorrhage in 236 consecutive cases in Brussels', *Stroke*, 1992, **23**, pp 24–7.

18 Stout, RW and Crawford, V, 'Seasonal variations in fibrinogen concentrations among elderly people', *Lancet*, 1991, **338**, pp 9–13.

19 Aylin, P *et al*, 'Temperature, housing, deprivation and their relationship to excess winter mortality in Great Britain, 1986–1996', *Int J Epidemiol*, 2001, **30**, pp 1100–8.

20 Kloner, RA *et al*, 'When throughout the year is coronary death most likely to occur? A 12-year population-based analysis of more than 220,000 cases', *Circulation*, 1999, **100**, pp 1630–34.

21 Rothwell, PM *et al*, 'Is stroke incidence related to season or temperature? The Oxfordshire Community Stroke Project', *Lancet*, 1996, **347**, pp 934–6: Sobel, E *et al* 'Stroke in the Lehigh Valley, seasonal variation in incidence rates', *Stroke*, 1987, **18**, pp 38–42.

22 Pasqualetti, P *et al*, 'Epidemiological chronorisk of stroke', *Acta Neurologica Scand*, 1990, **81**, pp 71–4.

23 Feigin, VL *et al*, 'A population-based study of the associations of stroke occurrence with weather parameters in Siberia, Russia (1982–92)', *Eur J Neurol*, 2000, **7**, pp 171–8.

24 Jakovljevic, D *et al*, 'Seasonal variation in the occurrence of stroke in a Finnish adult population. The FINMONICA Stroke Register. Finnish monitoring trends and determinants in cardiovascular disease', *Stroke*, 1996, **27**, pp 1774–9.

25 Tsementzis, SA *et al*, 'Seasonal variation of cerebrovascular disease', *Acta Neurochir (Wien)*, 1991, **111**, pp 80–83; Biller, J *et al*, 'Seasonal variation of stroke: does it exist?' *Neuroepidemiol*, 1988, **7**, pp 89–98; Azevedo, E *et al*, 'Cold: a risk factor for stroke?' *J Neurol*, 1995, **242**, pp 217–21.

26 Green, MS *et al*, 'Excess winter mortality from ischaemic heart disease and stroke during colder and warmer years in Israel', *Eur J Pub Health*, 1994, **4**, pp 3–11.

27 Shinkawa, A *et al*, 1990, *op. cit.*

28 Ricci, S *et al*, 'Diurnal and seasonal variations in the occurrence of stroke: a community-based study', *Neuroepidemiol*, 1992, **11**, pp 59–64.

29 Capon, A *et al*, 'Seasonal variation of cerebral haemorrhage in 236 consecutive cases in Brussels', *Stroke*, 1992, **23**, pp 24–7; Tsementzis, SA *et al*, 'Seasonal variation of cerebrovascular

diseases', *Acta Neurochirurgica*, 1991, **111**, pp 80–3.
30 Berginer, VM *et al*, 'Clustering of strokes in association with meteorologic factors in the Negev Desert of Israel: 1981–1983', *Stroke*, 1989, **20**, pp 65–69.
31 Lindsberg, PJ *et al*, 'Inflammation and infections as risk factors for ischemic stroke', *Stroke*, 2003, **34**, pp 2518–32.
32 Khaw, KT, 'Temperature and cardiovascular mortality', *Lancet*, 1995, **345**, pp 337–8.
33 Brennan, PJ *et al*, 'Seasonal variation in arterial blood pressure', *BMJ*, 1982, **285**, pp 919–23; Woodhouse, PR *et al*, 'Seasonal variation of blood pressure and its relationship to ambient temperature in an elderly population', *J Hypertens*, 1993, **11**, pp 1267–74; Izzo, JL *et al*, 'Hemodynamics of seasonal adaptation', *Am J Hyperten*, 1990, pp 405–7.
34 Brennan, PJ *et al*, 1982, *op. cit.*; Giaconi, S *et al*, 'Long-term reproducibility and evaluation of seasonal influences on blood pressure monitoring', *J Hypertens Suppl*, 1988, **6**, pp S64–6; Giaconi, S *et al*, 'Seasonal influences on blood pressure in high normal to mild hypertensive range', *Hypertension*, 1989, **14**, pp 22–7.
35 Stewart, S, Presentation given at the XXIV Congress of the European Society of Cardiology Annual meeting, 31 August – 4 September 2002. Abstract no. P1091.
36 Frost, L *et al*, 'Seasonal variation in hospital discharge diagnosis of atrial fibrillation: a population-based study', *Epidemiol*, 2002, **13**, pp 211–5.
37 Spengos, K *et al*, 'Diurnal and seasonal variation of stroke incidence in patients with cardioembolic stroke due to atrial fibrillation', *Neuroepidemiol*, 2003, **22**, pp 204–10.
38 Kristal-Boneh, E *et al*, 'Summer–winter differences in 24 h variability of heart rate', *J Cardiovasc Risk*, 2000, **7**, pp 141–6.
39 Keatinge, WR *et al*, 'Increases in platelet and red cell counts, blood viscosity, and arterial pressure during mild surface cooling: factors in mortality from coronary and cerebral thrombosis in winter', *BMJ*, 1984, **289**, pp 1405–8; Pasqualetti, P *et al*, 'Epidemiological chronorisk of stroke', *Acta Neurologica Scandinavica*, 1990, **81**, pp 71–4.
40 Stout, RW at Crawford, V, 'Seasonal variations in fibrinogen concentrations among elderly people', *Lancet*, 1991, **338**, pp 9–13.
41 Meade, TW *et al*, 'Hemostatic function and cardiovascular death: early results of a prospective study', *Lancet*, 1980, **17**, pp 1050–3;

Kannel, WB *et al*, 'Fibrinogen, cigarette smoking, and risk of cardiovascular disease insights from the Framingham Study', *Am Heart J*, 1987, **113**, pp 1006–10; Wilhelmsen, L *et al*, 'Fibrinogen as a risk factor for stroke and myocardial infarction', *N Engl J Med*, 1984, **311**, pp 501–5.

42 Woodhouse, PR *et al*, 'Seasonal variations of plasma fibrinogen and factor VII activity in the elderly: winter infections and death from cardiovascular disease', *Lancet*, 1994, **343**, pp 435–39; Stout, RW and Crawford, V, 1991, *op. cit.*

43 Garvey, L, 'The body barometer', *Health*, 1987, **19**, pp 80–5.

44 Stout RW *et al*, 'Seasonal changes in haemostatic factors in young and elderly subjects', *Age Aging*, 1996, **25**, pp 256–9.

45 Råstam, L *et al*, 'Seasonal variation in plasma cholesterol: implications for screening and referral', *Am J Prev Med*, 1992, **8**, pp 360–66; Woodhouse, PR *et al*, 'Seasonal variation of serum lipids in an elderly population', *Age Aging*, 1993, **22**, pp 273–8; Ockene, IS *et al*, 'Seasonal variation of cholesterol levels', presented at the 3rd International Conference on Preventive Cardiology, Oslo, Norway, June 1993; Van Gent, CM *et al*, 'High-density lipoprotein cholesterol, monthly variation and association with cardiovascular risk factors in 1000 forty-year-old Dutch citizens', *Clinica Chimica Acta*, 1978, **88**, pp 155–62; Gordon, DJ *et al*, 'Cyclic seasonal variation in plasma lipid and lipoprotein levels: the Lipid Research Clinics Coronary Primary Prevention Trial placebo group', *J Clin Epidemiol*, 1988, **41**, pp 679–89; Thomas, CB *et al*, 'Observations on seasonal variations in total serum cholesterol level among healthy young prisoners', 1961, **54**, pp 413–30; Fyfe T *et al*, 'Seasonal variation in serum lipids, and incidence and mortality of ischaemic heart disease', *J Atheroscler Res*, 1968, **8**, pp 591–96; Thelle, DS *et al*, 'The Tromsø Heart Study: Methods and main results of the cross-sectional study', *Acta Med Scand*, 1976, **200**, pp 107–18; Gordon, DJ, 'Cyclic seasonal variation in plasma lipid and lipoprotein levels: the Lipid Research Clinics Coronary Primary Prevention Trial Placebo Group', *J Clin Epidemiol*, 1988, **41**, pp 679–89; Buxtorf, JC *et al*, 'Seasonal variations of serum lipids and apoproteins', *Ann Nutr Metab*, 1988, **32**, pp 68–74; Cucu, F *et al*, 'Seasonal variations of serum cholesterol detected in the Bucharest Multifactorial Prevention Trial of Coronary Heart Disease – Ten years follow-up (1971–1982)', *Medecine Interne*, 1991, **29**, pp 15–21.

46 Keys, A *et al*, 'Serum-cholesterol studies in Finland', *Lancet*, 1958, **2**, pp 175–8.
47 Råstam, L *et al*, 'Seasonal variation in plasma cholesterol: Implications for screening and referral', *Am J Prev Med*, 1992, **8**, pp 360–66.
48 Gordon, DJ *et al*, 'Seasonal cholesterol cycles: the Lipid Research Clinics Coronary Primary Prevention Trial placebo group', 1987, **76**, pp 1224–31.
49 National Cholesterol Education Program, 'Report of the Expert Panel on detection, evaluation, and treatment of high blood cholesterol in adults', NIH Publication No. 88–2925, 1988, Expert Panel on detection evaluation and treatment of high blood cholesterol in adults; 'Summary of the second report of the National Cholesterol Education Program (NCEP) expert panel on detection, evaluation, and treatment of high blood cholesterol in adults (Adult Treatment Panel II)', *JAMA*, 1993, **269**, pp 3015–23.
50 Gordon, DJ *et al*, 'Cyclic seasonal variation in plasma lipid and lipoprotein levels: the Lipid Research Clinics Coronary Primary Prevention Trial Placebo Group', *J Clin Epidemiol*, 1988, **41**, pp 679–89.
51 Dobson, HM *et al*, 'The effect of ascorbic acid on the seasonal variation in serum cholesterol levels', *Scott Med J*, 1984, **29**, pp 176–82.
52 Mishmar, D *et al*, 'Natural selection shaped regional mtDNA variation in humans', *Proc Natl Acad Sci*, 2003, **100**, pp 171–76.
53 Uitenbroek, DG, 'Seasonal variation in leisure time physical activity', *Med Sci Sports Exerc*, 1992, **25**, pp 755–60.
54 Bergstralh, EJ *et al*, 'Effect of season on physical activity score, back extensor muscle strength, and lumbar bone mineral density', *J Bone Min Res*, 1990, **5**, pp 371–7.
55 Bluher, M *et al*, 'Influence of dietary intake and physical activity on annual rhythm of human blood cholesterol concentrations', *Chronobiol Int*, 2001, **18**, pp 541–57; Manttari, M *et al*, 'Seasonal variation in high-density lipoprotein cholesterol', *Atherosclerosis*, 1993, **100**, pp 257–65.
56 Gueldner, SH *et al*, 'Long-term exercise patterns and immune function in healthy older women. A report of preliminary findings', *Mech Ageing Dev*, 1997, **93**, pp 215–22.
57 Peters-Futre, EM, 'Vitamin C, neutrophil function, and upper respiratory tract infection risk in distance runners: the

missing link', *Exerc Immunol Rev*, 1997, **3**, pp 32–52; Brenner, IK *et al*, 'Infection in athletes', *Sports Med*, 1994, **17**, pp 86–107.
58 Nieman, DC, 'Exercise, infection, and immunity', *Int J Sports Med*, 1994, **15** (Suppl 3), pp S131–41.
59 Gaydos, JC *et al*, 'Swine influenza A at Fort Dix, New Jersey (January–February 1976). II, Transmission and morbidity in units with cases', *J Infect Dis*, 1977, **136** (Suppl), pp S363–8.
60 Spoont, MR *et al*, 'Dimensional measurement of seasonal variation in mood and behavior', *Psych Res*, 1991, **39**, pp 269–84.
61 Kräuchi, K and Wirz-Justice, A, 'The four seasons: food intake frequency in seasonal affective disorder in the course of a year', *Psychiatry Res*, 1988, **25**, pp 323–38.
62 Stump, B, 'Under the weather?', *Men's Health*, 1999, **14**, pp 124–41; Leviton, R, 1989, *op. cit.*
63 Sack, RL *et al*, 'Human melatonin production decreases with age', *J Pineal Res*, 1986, **3**, pp 379–88.
64 Rosen, LN *et al*, 'Prevalence of seasonal affective disorder at four latitudes', *Psychiatry Res*, 1990, **31**, pp 131–44.
65 Rosenthal, NE *et al*, 'Psychobiological effects of carbohydrate- and protein-rich meals in patients with seasonal affective disorder and normal controls', *Biol Psychiatry*, 1989, **25**, pp 1029–40.
66 Lewy, AJ *et al*, 'Morning vs evening light treatment of patients with winter depression', *Arch Gen Psychiatry*, 1998, **55** (10), pp 890–6.
67 de Castro, JM, 'Seasonal rhythms of human nutrient intake and meal pattern', *Physiol Behav*, 1991, **50**, pp 243–8.
68 Spoont, MR *et al*, 1991, *op. cit.*
69 McGrath, J, 'Hypothesis: is low prenatal vitamin D a risk-modify ing factor for schizophrenia?' *Schizophrenia Res*, 1999, **40**, pp 173–177; Welham, J *et al*, 'Climate, geography and the search for candidate non-genetic risk factors for schizophrenia', *Int J Mental Health*, 2000, **29**, pp 79–100; de Messias, E *et al*, 'Schizophrenia and season of birth in a tropical region, relationship to rainfall', *Schizophrenia Res*, 2001, **48**, p 227; McGrath, J *et al*, 'The impact of low prenatal vitamin D on brain development: using an animal model to examine the vitamin D hypothesis of schizophrenia', *Schizophr Res*, 2001, **49**, p 48.
70 Hajat, S *et al*, 'Association of air pollution with daily GP consultations for asthma and other lower respiratory conditions in London', *Thorax*, 1999, **54**, pp 597–605.

71 Cohen, S *et al*, 'Smoking, alcohol consumption, and susceptibility to the common cold', *Am J Public Health*, 1993, **83**, pp 1277–83.
72 Klein, TW, 'Stress and infections', *J Fla Med Assoc*, 1993, **80**, pp 409–11; Drummond, PD and Hewson-Bower, B 'Increased psychosocial stress and decreased mucosal immunity in children with recurrent upper respiratory tract infections', *J Psychosom Res*, 1997, **43**, pp 271–8.
73 Cohen, S *et al*, 'Types of stressors that increase susceptibility to the common cold in healthy adults', *Health Psychol*, 1998, **17**, pp 214–23.
74 Cohen, S *et al*, 'Psychological stress and susceptibility to the common cold', *N Engl J Med*, 1991, **325**, pp 606–12.
75 Cohen, S *et al*, 'Social ties and susceptibility to the common cold', *JAMA*, 1997, **277**, pp 1940–4.
76 Takkouche, B *et al*, 'A cohort study of stress and the common cold', *Epidemiol*, 2001, **12**, pp 345–9.
77 Kilpeläinen, M *et al*, 'Home dampness, current allergic diseases, and respiratory infections among young adults', *Thorax*, 2001, **56**, pp 462–67.
78 Brunekreef, B, 'Damp housing and adult respiratory symptoms', *Allergy*, 1992, **47**, pp 498–502; Dales, RE *et al*, 'Adverse health effects among adults exposed to home dampness and molds', *Am Rev Respir Dis*, 1991, **143**, pp 505–9.
79 Ronmark, E *et al*, 'Asthma, type-1 allergy and related conditions in 7- and 8-year-old children in northern Sweden: prevalence rates and risk factor pattern', *Respir Med*, 1998, **92**, pp 316–24; Andriessen, JW *et al*, 'Home dampness and respiratory health status in European children', *Clin Exp Allergy*, 1998, **28**, pp 1191–200.
80 Cariñanos, P *et al*, 'Meteorological phenomena affecting the presence of solid particle suspended in the air during winter', *Int J Biometeorol*, 2000, **1**, pp 6–10.
81 Mellion, MB and Kobayashi, RH, 'Exercise-induced asthma', *Am Fam Physician*, 1992, **45**, pp 2671–7.
82 Spector, SL, 'Update on exercise-induced asthma', *Ann Allergy*, 1993, **71**, pp 571–7.
83 Brusasco, V and Crimi, E, 'Allergy and sports: exercise-induced asthma', *Int J Sports Med*, 1994, **15** (Suppl 3), pp S184–6; Eggleston, PA, 'Pathophysiology of exercise-induced asthma', *Med Sci Sports Exerc*, 1986, **18**, pp 318–21.
84 Pierson, WE *et al*, 'Implications of air pollution effects on athletic performance', *Med Sci Sports Exerc*, 1986, **18**, pp 322–7;

Pierson, WE *et al*, 'Air pollutants, bronchial hyperreactivity, and exercise', *J Allergy Clin Immunol*, 1984, **73**, pp 717–21.

85 Garvey, L, 1987, *op. cit.*

86 Rietveld, WJ *et al*, 'Seasonal fluctuations in the cervical smear rates for (pre) malignant changes and for infections', *Diagnostic Cytopathology*, 1997, **17**, pp 452–5.

87 Hermida, RC and Ayala, DE, 'Reproducible and predictable yearly pattern in the incidence of uterine cervical cancer', *Chronobiol Int*, 1996, **13**, pp 305–16.

88 Grant, WB, 'An estimate of premature cancer mortality in the US due to inadequate doses of solar ultraviolet-B radiation, Cancer', 2002, **94**, pp 1867–75.

89 Mason, BH and Holdaway, IM, 'The seasonal variation in breast cancer detection: its significance and possible mechanisms', *J Royal Soc N Zeal*, 1994, **24**, pp 439–49.

90 Angwafo, FF, 'Migration and prostate cancer; an international perspective', *J Natl Med Assoc*, 1998, **90** (11 suppl), pp SL20–3.

91 Page, EH and Shear, NH, 'Temperature-dependent skin disorders', *J Am Acad Dermatol*, 1988, **18**, pp 1003–19.

92 Centers for Disease Control 'Exposure-related hypothermia deaths – District of Columbia 1972–1982', *Morb Mortal Wkly Rep*, December 1982, pp 31–50.

93 Fitzgerald, FT and Jessop, C, 'Accidental hypothermia: a report of 22 cases and review of the literature', *Adv Intern Med*, 1982, **27**, pp 127–150; Lewin, S *et al*, 'Infections in hypothermia inpatients' *Arch Intern Med*, 1981, **141**, pp 920–25; Bristow, GR *et al*, 'Resuscitation from cardiopulmonary arrest during accidental hypothermia due to exhaustion and exposure', *Can Med Assoc J*, 1977, **117**, pp 247–8; Hudson, LD and Conn, RD, 'Accidental hypothermia: associated diagnoses and prognosis in a common problem', *JAMA*, 1974, **227**, pp 37–40.

94 Collins, RJ *et al*, 'Accidental hypothermia and impaired temperature homeostasis in the elderly', *BMJ*, 1977, **1**, pp 353–6; Collins, RJ *et al*, 'Shivering thermogenesis and vasomotor responses with convective cooling in the elderly', *J Physiol*, 1981, **320**, p 76; Wagner, JA *et al*, 'Age and temperature regulation of humans in neutral and cold environments', *J Appl Physiol*, 1974, **37**, pp 562–5.

95 Collins, KJ *et al*, 'Urban hypothermia: preferred temperature and thermal perception in old age', *BMJ*, 1981, **282**, pp 175–7.

96 Rango, NR, 'Exposure-related hypothermia mortality in the United States, 1970–1979', *Am J Publ Health*, 1984, **74**, pp

1159–60; Centers for Disease Control, 1982, *op. cit.*

97 Cunningham, DJ *et al*, 'Comparative thermoregulatory responses of resting men and women', *Journal of Applied Physiology*, 1978, **45**, pp 908–15; Hardy, JD and DuBois, EF, 'Differences between men and women in their response to heat and cold', *Proc Natl Acad Sci*, 1940, **26**, pp 389–398; Graham, TE, 'Alcohol ingestion and sex differences on the thermal responses to mild exercise in a cold environment', *Human Biol*, 1983, **55**, pp 463–476; Wyndham, CH *et al*, 'Physiological reactions to cold of Caucasion females', *J Appl Physiol*, 1964, **19**, pp 877–80.

98 Bernstein, LM *et al*, 'Body composition as related to heat regulation in women', *J Appl Physiol*, 1956, **9**, pp 241–56; Gallow, D *et al*, 'Comparative thermoregulatory responses to acute cold in women of Asian and European descent', *Human Biol*, 1984, **56**, pp 19–34; Veicsteinas, A *et al*, 'Superficial shell insulation in resting and exercising men in water', *J Appl Physiol*, 1982, **52**, pp 1557–64.

99 Graham, TE and Lougheed, MD, 'Thermal responses to exercise in the cold: influence of sex difference and alcohol', *Human Biol*, 1985, **57**, pp 687–98.

100 Kalkstein, 1984, *op. cit.*

101 Foray, J, 'Mountain frostbite: current trends in prognosis and treatment (from results concerning 1261 cases)', *Int J Sports Med*, 1992, **13** (Suppl 1), pp S193–6.

102 Pulla, RJ *et al*, 'Frostbite: an overview with case presentations', *J Foot Ankle Surg*, 1994, **33**, pp 53–63; Antti-Poika, I, 'Severe frostbite of the upper extremities – a psychosocial problem mostly associated with alcohol abuse', *Scand J Soc Med*, 1990, **18**, pp 59–61.

103 Rosen, L *et al*, 'Local cold injuries sustained during military service in the Norwegian Army', *Arctic Med Res*, 1991, **50**, pp 159–65; Hermann, G *et al* 'The problem of frostbite in civilian medical practice', *Surg Clin North Am*, 1963, **43**, pp 519–36; Ervasti, E, 'Frostbites of the extremities and their sequelae', *Acta Chir Scand Suppl*, 1962, **299**, pp 1–69.

104 Valnicek, SM *et al*, 'Frostbite in the prairies: a 12–year review', *Plast Reconstr Surg*, 1993, **92**, pp 633–41.

105 Bracker, MD, 'Environmental and thermal injury', *Clin Sports Med*, 1992, **11**, pp 419–36; Christenson, C and Stewart, C, 'Frostbite', *Am Fam Physician*, 1984, **30**, pp 111–22; Boswick, JA *et al*, 'The epidemiology of cold injuries', *Surg Gynecol Obstet*, 1979, **149**, pp 326–32; Kyosola, K, 'Clinical

experiences in the management of cold injuries: a study of 110 cases', *J Trauma*, 1974, **14**, pp 32–6.

106 McCauley, RL *et al*, 'Frostbite and other cold-induced injuries', in Auerbach, PS (ed), *Wilderness Medicine*, 3rd edn, St Louis, Mosby, 1995.

## Chapter 8: Artificial Environments

1 Wallace, LA *et al*, *Personal Exposure, Indoor–outdoor Relationships, and Breath Levels of Toxic Air Pollutants Measured for 355 Persons in New Jersey*, EPA 0589, and Wallace, LA *et al*, *Personal Exposures, Outdoor Concentrations, and Breath Levels of Toxic Air Pollutants Measured for 425 persons in Urban, Suburban and Rural Areas*, EPA 0589, presented at annual meeting of Air Pollution Control Association, San Francisco, CA, 25 June 25 1984; Ott, WR and Roberts, JW, 'Everyday exposure to toxic pollutants', *Scientific American*, Feb 1998, pp 86–91.

2 Hill, RH Jr *et al*, 'p–Dichlorobenzene exposure among 1,000 adults in the United States', *Arch Environ Health*, 1995, **50**, pp 277–80.

3 EPA Office of Toxic Substances, *Broad Scan Analysis of the FY82 National Human Adipose Tissue Survey Specimens*, EPA 560/5-86-035, Springfield, VA: National Technical Information Service (NTIS), No. PB 87–177218/REB, 1982.

4 *Indoor Air Quality in the Home, Nitrogen Dioxide, Formaldehyde, Volatile Organic Compounds, House Dust Mites, Fungi and Bacteria* (Assessment A2), Leicester, UK: Institute for Environment and Health, 1996.

5 Brown, SK, 'Exposure to volatile organic compounds in indoor air: a review', in *Proceedings of the 11th International Clean Air Conference of the Clean Air Society of Australia and New Zealand*, Brisbane: Clean Air Society of Australia and New Zealand, 1992, **1**, pp 95–104; Brown, SK, 'Volatile organic pollutants in new and established buildings in Melbourne, Australia', *Indoor Air*, 2002, **12**, pp 55–63.

6 Joint Research Committee, 'Human exposure to indoor air pollution: do you really know what you are breathing when sitting at home?', *JRC/ISPRA*, September 2003.

7 Middaugh, DA *et al*, 'Sick building syndrome: medical evaluation of two work forces', *J Occup Med*, 1992, **34**, pp 1197–1203; World Health Organization, *Indoor Air Pollutants, Exposure and Health Effects*, EURO Reports and Studies 78, World Health

Organization, 1983; Morris, L and Hawkins, L, 'The role of stress in the sick building syndrome', in Siefert B *et al* (eds), *Indoor Air '87, Proceedings of the 4th International Conference on Indoor Air Quality and Climate, Berlin (West), Institute for Water, Soil and Air Hygiene*, 1987, **2**, pp 566–571; Wilson, S and Hedge, A, *The Office Environment Survey: A Study of Building Sickness*, London: Building Use Studies Ltd, 1987.

8 Rogers, SA, 'Diagnosing the tight building syndrome', *Environ Health Perspect*, 1987, **76**, pp 195–198; Menzies, R *et al*, 'Impact of exposure to multiple contaminants on symptoms of sick building syndrome', *Proc Indoor Air*, 1993, **1**, pp 363–68.

9 EPA, *Healthy buildings, healthy people: a vision for the 21st century*, US Environmental Protection Agency, Office of Air and Radiation, EPA Report 402-K-00-002, March 2000.

10 Thomas, P, *Living Dangerously: Are Everyday Toxins Making You Sick?*, Dublin: New Leaf, 2003.

11 He, J *et al*, 'Passive smoking and the risk of coronary heart disease, a meta-analysis of epidemiologic studies', *N Engl J Med*, 1999, **340**, pp 920–6; Kawachi, I *et al*, 'A prospective study of passive smoking and coronary heart disease', *Circulation*, 1997, **95**, pp 2374–79; Wells, AJ, 'Heart disease from passive smoking in the workplace', *J Am Coll Cardiol*, 1998, **31**, p 19; Muscat, JE and Wynder, EL, 'Exposure to environmental tobacco smoke and the risk of heart attack', *Int J Epidemiol*, 1995, **24**, p 7159; Panagiotakos, DB *et al*, 'The association between secondhand smoke and the risk of developing acute coronary syndromes, among non–smokers, under the presence of several cardiovascular risk factors: the CARDIO2000 case-control study', *BMC Public Health*, 2002, **2**, p 9.

12 Howard, G and Thun, MJ, 'Why is environmental tobacco smoke more strongly associated with coronary heart disease than expected? A review of potential biases and experimental data', *Environ Health Perspect*, 1999, **107** (Suppl 6), pp 853–8.

13 Cobb, N and Etzel, RA, 'Unintentional carbon monoxide-related deaths in the United States, 1979 through 1988', *JAMA*, 1991, **266**, pp 659–63.

14 Gould, D, 'Legionnaires' disease', *Nurs Stand*, 2003, **17**, pp 41–4; McEvoy, M *et al* 'A cluster of cases of legionnaires' disease associated with exposure to a spa pool on display', *Commun Dis Public Health*, 2000, **3**, pp 43–5.

15 EPA *et al*, *Indoor Air Pollution: An Introduction for Health*

*Professionals*, Environmental Protection Agency, US Government Printing Office Publication No. 1994–523–217/81322, 1994.

16 Kilbourne EM *et al*, 'Risk factors for heatstroke, a case-control study', *JAMA*, 1982, **247**, pp 3332–6; Marmor, M, 'Heatwave mortality in New York City, 1949 to 1970', *Arch Environ Health*, 1975, **30**, pp 131–6.

17 Curwen, M, 'Excess winter mortality in England and Wales with special reference to the effects of temperature and influenza' in Charlton, J and Murphy, M (eds), *The Health of Adult Britain 1841–1994, Volume 1,* London: HMSO, 1997.

18 Keatinge, WR, 'Seasonal mortality among elderly people with unrestricted home heating', *BMJ* (Clinical Research Ed), 1986, **293**, pp 732–3.

19 Parsons, AG, 'The association between daily weather and daily shopping patterns', *Austalasian Marketing J*, 2001, **9**, pp 78–83.

20 Hedge, A, 'Suggestive evidence for a relationship between office design and self-reports of ill-health among office workers in the United Kingdom', *J Architect Plan Res*, 1984, **1**, pp 163–74; work and environment measurements in two office buildings with different ventilation systems', *BMJ*, 1985, **291**, pp 373–76; Burge, PS *et al*, 'Sick building syndrome: a study of 4373 office workers', *Ann Occup Hygiene*, 1987, **31**, pp 493–504; Hedge, A, 'Environmental conditions and health in offices', *Int Rev Ergonomics*, 1989, **3**, pp 87–110; Mendell, M and Smith, A, 'Consistent pattern of elevated symptoms in airconditioned office buildings: a reanalysis of epidemiologic studies', *Am J Pub Health*, 1990, **80**, pp 1193–99; Zweers, T *et al*, 'Health and indoor climate complaints of 7043 office workers in 61 buildings in the Netherlands', *Indoor Air*, 1992, **2**, pp 127–36; Mendell, MJ, 'Non-specific symptoms in office workers: a review and summary of the epidemiologic literature', *Indoor Air*, 1993, **3**, pp 227–36.

21 Fang, LG *et al*, 'Impact of temperature and humidity on the perception of indoor air quality', *Indoor Air*, 1998, **8**, pp 80–90; Fang, LG *et al*, 'Impact of temperature and humidity on perception of indoor air quality during immediate and longer whole-body exposures', *Indoor Air*, 1998, **8**, pp 276–84; Fang, LG *et al*, 'Field study on the impact of temperature, humidity and ventilation on perceived air quality', in *Indoor Air 99. The Eighth International Conference*

*on Indoor Air Quality and Climate*, 1999, **2**, pp 107–12.

22 Engvall, K *et al*, 'Ocular, nasal, dermal and respiratory symptoms in relation to heating, ventilation, energy conservation, and reconstruction of older multi-family houses', *Indoor Air*, 2003, **13**, p 206.

23 *Asthma and the Environment, A Strategy to Protect Children*, President's Task Force on Environmental Health Risks and Safety Risks to Children, 28 January 1999.

24 Myhrvold, AN *et al*, 'Indoor environment in schools – pupils health and performance in regard to $CO_2$ concentrations', in *Indoor Air '96. The Seventh International Conference on Indoor Air Quality and Climate*, 1996, **4**, pp 369–371.

25 Wyon, DP *et al*, 'The effects of moderate heat stress on mental performance', *Scand J Work Environ Health*, 1979, **5**, pp 352–61; Wyon, DP, 'The ergonomics of healthy buildings, overcoming barriers to productivity', in *IAQ '91, Post Conference Proceedings*, American Society of Heating, Refrigerating, and Air-Conditioning Engineers, Atlanta, 1991, pp 43–46.

26 Wargocki, P *et al*, 'Perceived air quality, SBS–symptoms and productivity in an office at two pollution loads' in *Indoor Air '99. The Eighth International Conference on Indoor Air Quality and Climate*, 1999, **2**, pp 131–6.

27 Repace, J *et al*, 'Fact Sheet On Secondhand Smoke', Internet Review Paper prepared for GlobaLink, an internet service of the UICC (International Union Against Cancer), Switzerland, Geneva, 23 February 1999.

28 Lysne, HN *et al*, 'Hygienic conditions in ventilation systems and the possible impact on indoor air microbial flora', in *Proceedings of Indoor Air '99, Edinburgh, 8th International Conference on Indoor Air Quality and Climate*, 1999, **2**, pp 220–4.

29 Morey, PR and Williams, CM, 'Is porous insulation material inside an HVAC system compatible with a healthy building?', in *Proceedings of IAQ '91 Healthy Buildings*, American Society of Heating, Refrigerating, and Air-Conditioning Engineers, Atlanta, 1991, pp 128–141.

30 Ott, JN, 'Some observations on the effect of the pigment epithelial cells of the retina of a rabbit's eye: recent progress in photobiology', in *Proceedings of the 4th International Congress on Photobiology*, Oxford: Blackwell Publications, July 1964, pp 395–396.

31 Ott, JN, 'School lighting and hyperactivity', *J Biosoc Res*, Summer

1980, **8**, pp 6–7; Ott, JN, 'Influence of fluorescent lights on hyperactivity and learning disabilities', *J Learning Dis*, August–September 1976, **9**, pp 417–22; Ott, JN *et al*, 'Light radiation and academic achievement: second year data', *Academic Ther*, Summer 1976, **4**, pp 397-407.

32 Wohlfarth, H and Sam, C, 'The effect of color psychodynamic environmental modification upon psychophysiological and behavioral reactions of severely handicapped children', *Int J Biosocial Res*, 1982, **3**, pp 10–38.

33 Hollwich, F and Dieckhues, B, 'The effect of natural and artificial light via the eye on the hormonal and metabolic balance of animal and man', *Opthalmologica*, 1980, **180**, pp 188–97.

34 Sharon, I *et al*, 'The effects of lights of different spectra on caries incidence in the golden hamster', *Arch Oral Biol*, 1971, **16**, pp 1427–31.

35 Holick, MF *et al*, 'Photosynthesis of previtamin D-3 in human skin and the physiologic consequences', *Science*, 1980, **210**, pp 203–5.

36 Beral, V *et al*, 'Malignant melanoma and exposure to fluorescent light at work', *Lancet*, 1982, **2**, pp 290–2.

37 Lytle, CD *et al*, 'An estimation of squamous cell carcinoma risk from ultraviolet radiation emitted by fluorescent lamps', *Photodermatol Photoimmunol Photomed*, 1992–93, **9**, pp 268–74; Wiskemann, A *et al*, 'Fluorescent lighting enhances chemically induced papilloma formation and increases susceptibility to tumor challenge in mice', *J Cancer Res Clin Oncol*, 1986, **112**, pp 141–3.

38 *Ultra-Violet Radiation Exposure From Tungsten Halogen Light Sources*, Health and Safety Executive / Local Authorities Enforcement Liaison Committee (HELA), HELA Data Sheet HSE 55916, September 2000.

39 Smith-Sonneborn, J, 'DNA repair and longevity assurance in paramecium Tetraurelia', *Science*, 1979, **203**, pp 1115–7; Smith-Sonneborn, J, 'Aging in protozoa', *Rev Biol Res Aging*, 1983, **1**, pp 29–35.

40 Liberman, J, *Light:, Medicine of the Future*, Santa Fe, NM: Bear, 1991.

41 Monk, TH and Folkard, S, *Making Shiftwork Tolerable*, London: Taylor and Francis, 1992.

42 'Special Issue: Night and Shiftwork', *Ergonomics*, **36**, Jan–Mar 1993.

43 Smoyer, KE *et al*, 'Heat-stress-related mortality in five cities in Southern Ontario, 1980–1996', *Int J Biometeorol*, 2000, **44**, pp 190–7; Kalkstein, LS *et al*, *The Impact of Cclimate on Canadian*

*Mortality, Present Relationships and Future Scenarios*, Report No 93–7, Downsview, Environment Canada/Canadian Climate Centre, 1993; Tavares, D, 'Weather and heat-related morbidity relationships in Toronto (1979–1989)', in Mortsch, L and Mills, B (eds), *Great Lakes – St Lawrence Basin Project Progress Report No. 1: Adapting to the Impacts of Climate Change and Variability*, Downsview, Environment Canada, 1996.

44 Smoyer, KE *et al*, The impacts of weather and pollution on human mortality in Birmingham, Alabama and Philadelphia, Pennsylvania', *Int J Climate*, 2000, **20**, pp 881–97; Macey, SM and Schneider, DF, 'Deaths from excessive heat and excessive cold among the elderly', *Gerontologist*, 1993, **33**, pp 497–500; Mackenbach, JP and Borst, V, 'Health-related mortality among nursing-home patients', *Lancet*, 1997, **349**, pp 1297–8.

45 Semenza, JC *et al*, Heat-related deaths during the July 1996 heat wave in Chicago', *New England Journal of Medicine*, 1996, **335**, pp 84–90; Kilbourne, EM *et al*, 1982, *op cit.*

46 Enquselassie, F *et al*, 'Seasons, temperature and coronary disease', *Int J Epidemiol*, 1993, **22**, pp 632–6; Auliciems, A and Frost, D, 'Temperature and cardiovascular deaths in Montreal', *Int J Biometeorol*, 1989, **33**, pp 151–6; Khaw, KT, 'Temperature and cardiovascular mortality', *Lancet*, 1995, **345**, pp 337–8.

47 Centers for Disease Control, 'Heat-related mortality', *Chicago*, July 1995, MMWR 1995, **44**, pp 577–9; Kalkstein, LS and Davis, RE, 'Weather and human mortality: an evaluation of demographic and interregional responses in the United States', *Ann Assoc Am Geographers*, 1989, **79**, pp 44–64.

48 Lutgens, FK and Tarbuck, EJ, *Atmosphere: Introduction To Meteorology*, Englewood Cliffs, NJ: Prentice-Hall, 1982; Changnon, SA 'Urban effects on severe local storms at St Louis', *J Appl Meteorol*, 1978, **17**, pp 578–86; Jáuregui, E and Luyando, E, 'Global radiation attenuation by air pollution and its effects on the thermal climate in Mexico City region', *Int J Climatol*, 1999, **19**, pp 683–94.

49 Burnett, RT *et al*, 'The effect of the urban ambient air pollution mix on daily mortality rates in 11 Canadian cities', *Can J Pub Health*, 1998, **89**, pp 152–6.

50 Burnett, RT *et al*, 'The association between ambient carbon monoxide levels and daily mortality in Toronto, Canada', *J Air Waste Management Assoc*, 1998, **48**, pp 689–700.

51 Salinas, M and Jeanette, V, 'The effect of outdoor air pollution on mortality risk: an ecological study from Santiago, Chile',

*Rapport Trimestriel de Statistique Sanitaire Mondaile*, 1995, **48**, pp 118–25; Ostro, B *et al*, 'Air pollution and mortality: results from a study of Santiago, Chile', *J Expos Anal Environ Epidemiol*, 1996, **6**, pp 97–114.

52 Sunyer J *et al*, 'Air pollution and mortality in Barcelona', *J Epidemiol Commun Health*, 1996, **50** (Suppl 1), pp S76–S80; Zmirou, D *et al*, 'Short-term effects of air pollution on mortality in the city of Lyon, France, 1985–90', *J Epidemiol Commun Health*, 1996, **50** (Suppl 1), pp S30–S35; Eilers, P and Groot, B, 'Effects of ambient particulate matter and ozone on daily mortality in Rotterdam, The Netherlands', *Arch Environ Health*, 1997, **52**, pp 455–64.

53 Xu, Z *et al*, 'Air pollution and daily mortality in Shenyang, China', *Arch Environ Health*, 2000, **55**, pp 115–20.

54 Borja-Aburto, VH *et al*, 'Mortality and ambient fine particles in southwest Mexico City, 1993–1995', *Environ Health Perspect*, 1998, **106**, pp 849–55.

55 Jendritzky, G and Bucher, K, 'Medical-meteorological funda mentals and their utilization in Germany', in Maarouf, A (ed), *Proceedings of The Weather and Health Workshop, Ottawa*, Environment Canada and Health Canada, November 1992, pp 42–59.

56 Lebowitz, MD *et al*, 'Health and the urban environment. XV. Acute respiratory episodes as reaction by sensitive individuals to air pollution and weather', *Environ Res*, 1973, **5**, pp 135–41.

57 Katsouyanni, K *et al*, 'Evidence for interaction between air pollution and high temperature in the causation of excess mortality', *Arch Environ Health*, 1993, **48**, pp 235–42.

58 Choi, K *et al*, 'Air pollution, temperature, and regional differences in lung cancer mortality in Japan', *Arch Environ Health*, 1997, **52**, pp 160–8.

59 Oke, TR, *Boundary Layer Climates*, New York: Cambridge University Press, 1987.

60 Shepherd, JM and Burian, SJ, 'Detection of urban-induced rainfall anomalies in a major coastal city', *Earth Interactions*, 2003, **7**, pp 1–14.

61 Changnon, S, 'Urban effects on freezing rain occurrences', *J Appl Meteorol*, 2003, pp 863–70.

62 Gribbin, J and Gribbin, M, *Ice Age – How a Change of Climate Made Us Human*, London: Penguin, 2001.

63 IPCC, Summary for Policymakers, Third Assessment Report, March 2001.

64 Nicholls, N *et al*, 'Observed climate variability and change', in

Houghton, JT *et al* (eds), *Climate Change 1995: The Science of Climate Change*, Cambridge: Cambridge University Press, 1996, pp 133–92.

65 NASA, *Global Temperature Trends: 1998 Global Surface Temperature Smashes Record*, NASA Goddard Institute for Space Studies, 16 December 1998; accessed at: www.giss.nasa.gov/research/ob serve/surftemp/

66 McMichael, AJ and Githeko, A, 'Human health', in McCarthy, JJ *et al* (eds), *Climate Change 2001: Impacts, Adaptation And Vulnerability*, Contribution of Working Group II to the Third Assessment Report of the Intergovernmental Panel on Climate Change, Cambridge: Cambridge University Press, 2001, pp 451–485.

67 Cox, PM *et al*, 'Acceleration of global warming due to carbon-cycle feedbacks in a coupled climate model', *Nature*, 2000, **408**, pp 184–7.

68 Smith, R, 'Doctors and climate change: action is needed because of the high probability of serious harm to health', *BMJ*, 1994, **309**, pp 1384–6.

69 Editorial, 'Climate change: the new bioterrorism', *Lancet*, 2001, **358**, p 1657; Kunzli, N, 'Public-health impact of outdoor and traffic-related air pollution: a European assessment', *Lancet*, 2000, **356**, pp 795–801.

70 LaPorte, RA *et al*, 'Health and climate change (Letter)', *Lancet*, 1994, **343**, pp 302–3; Epstein, P 'Climate and health', *Science*, 1999, **285**, pp 347–8.

71 McMichael, AJ, *Human Frontiers, Environments and Disease: Past Patterns, Uncertain Futures*, Cambridge: Cambridge University Press, 2001.

72 Moos, RH, *The Human Context: Environmental Determinants of Behavior*, New York: John Wiley and Sons, 1976; Garvey, L, 'The body barometer', *Health*, 1987, **19**, pp 80–5.

73 Kovats, RS *et al*, 'Climate change and human health in Europe', *BMJ*, 1999, **318**, pp 1682–5; Caplan, CE, 'The big chill: diseases exacerbated by exposure to cold', *CMAJ*, 1999, **160**, p 33.

74 Karl, TR and Knight, RW, 'The 1995 Chicago heatwave: how likely is a recurrence?' *Bull Am Meteorological Soc*, 1997, **78**, pp 1107–19.

75 Kalstein, LS and Greene, JS, 'An evaluation of climate/mortality in large US cities and the possible impact of a climate change', *Environ Health Perspect*, 1997, **105**, pp 84–93.

76 McMichael, AJ, 'Human population health', in Watson, RT *et al*

(eds), *Climate Change 1995 – Impacts, Adaptations, and Mitigation of Climate Change: Scientific-Technical Analyses*, Cambridge: Cambridge University Press, 1996, pp 561–84.

77 ven der Leun, JC and De Gruijl, FR, 'Climate change and skin cancer', *Photochem Photoboll Sci*, 2002, **1**, pp 324–6; De Gruijl, FR *et al* 'Skin cancer and solar radiation', *Eur J Cancer*, 1999, **35**, pp 2003–9.

78 Albritton, DL *et al*, 'Climate change 2001: the scientific basis', in *Intergovernmental Panel on Climate Change (IPCC) Working Group I. Summary for policy makers. Third assessment report*. New York: Cambridge University Press, 2001; National Research Council, National Academy of Sciences, *Abrupt Climate Change: Inevitable Surprises*, Washington, DC: National Academy Press, 2001.

79 Kattenberg, AF *et al*, 'Climate models – projections of future climate', in Houghton' JT *et al* (eds), *Climate Change 1995: The Science of Climate Change*, Cambridge: Cambridge University Press, 1996, pp 285–357.

80 Karl, TR *et al*, 'Quayle, Indices of climate change for the United States', *Bull Am Meteorological Soc*, 1996, **77**, pp 279–92.

81 Groisman, PY and Easterling, DR, 'Variability and trends of precipitation and snowfall over the United States and Canada', *J Climate*, 1994, **7**, pp 184–205.

82 Karl, TR and Knight, RW, 'Secular trends of precipitation amount, frequency, and intensity in the United States', *Bull Am Meteorological Soc*, 1998, **79**, pp 231–41.

83 McMichael, AJ *et al* (eds), *Climate Change and Human Health*, World Health Organization, World Meteorological Organization, United Nations Environmental Program, Switzerland, Geneva, 1996; Epstein, PR, 'Climate and health', *Science*, 1999, **285**, pp 347–8; Epstein, PR *et al*, 'Biological and physical signs of climate change, focus on mosquito borne diseases', *Bull Am Meteorological Soc*, 1998, **79**, pp 409–17.

84 CDC, *West Nile-like Virus in the United States*, Centers for Disease Control and Prevention, Updated 5 October 1999.

85 CDC, *Questions and Answers about West-Nile Encephalitis, Division of Vector-Borne Infectious Diseases*, National Center for Infectious Diseases, Centers for Disease Control and Prevention, Revised 24 November 1999.

86 Parmenter, RR *et al* 'Incidence of plague associated with increased winter–spring precipitation in New Mexico', *Am J Trop Med Hygiene*, 1999, **61**, pp 814–21.

87 Lindgren, E and Gustafson, R, 'Tick-borne encephalitis in Sweden and climate change', *Lancet*, 2001, **358**, pp 16–8.
88 Patz, JA *et al*, 'Climate change: regional warming and malaria resurgence', *Nature*, 2002, **420**, pp 627–8.
89 Gregory, JM *et al*, 'Summer drought in northern midlatitudes in a time-dependent CO2 climate experiment', *J Climate*, 1997, **10**, pp 662–86.
90 Dilley, M and Heyman, B, 'ENSO and disaster: droughts, floods, and El Niño/Southern Oscillation warm events', *Disasters*, 1995, **19**, pp 181–93.
91 Bourma, MJ *et al*, 'Global assessment of El Niño's disaster burden', *Lancet*, 1999, **350**, pp 1435–8.
92 Kwok, R and Comiso, JC, 'Southern ocean climate and sea ice anomalies associated with the Southern Oscillation', *J Climate*, 2002, **15**, pp 487–501.
93 WHO, *El Niño and its Health Impacts*, Geneva: World Health Organization, Fact Sheet No. 192, 2000.

# Bibliography


Allan, TM and Douglas, AS, *Seasonal Variations in Health and Disease – A Bibliography*, London: Mansell, 1994.

Becker, R, *The Body Electric: Electromagnetism and the Foundation of Life*, New York: Quill, 1985.

Durschmied, E, *The Weather Factor*, London: Coronet, 2000.

Fast, J, *Weather Language – How Climate Affects your Body and Mind…and What to Do About It*, New York: Wyden, 1979.

Folk, GE, *Textbook of Environmental Physiology*, Philadelphia, PA: Lee and Febiger, 1974.

Freier, GD, *Weather Proverbs*, Tuscon, AZ: Fisher Books, 1992.

Garvey, L, 'The Body Barometer', *Health*, 1987 **19**, pp 80–5.

Gribbin, J and Gribbin, M, *Ice Age – How a Change of Climate Made Us Human*, London: Penguin, 2001.

Haggerty, D, *Rhymes to Predict the Weather*, Seattle, WA: Spring Meadow Publishers, 1985.

Henson, R, *The Rough Guide to the Weather*, London: Rough Guides Ltd, 2002.

Huntington, E, *Civilization and Climate*, 3rd edn, New Haven: Yale University Press, 1924.

Hyman, JW, *The Light Book: How Natural And Artificial Light Affect Our Health, Mood, And Behavior*, Los Angeles: Jeremy P Tarcher, 1990.

Kaiser, M, *How the Weather Affects your Health*, Melbourne: Michelle Anderson, 2002.

Landsberg, HE, *Weather and Health: An Introduction to Biometeorology*, New York: Doubleday/Anchor, 1969.

Licht, S (ed), *Medical Climatology*, Baltimore, MD: Waverly Press, 1964.

Lieber, AL, *How the Moon Affects You*, Mamaroneck, NY: Hastings House, 1996.

Ludlum, DM (ed), *Weather*, London: HarperCollins, 2001.

Morgan, MD and Moran, JM, *Weather and People*, New Jersey: Prentice-Hall, 1997.

Ott, J, *Light, Radiation and You: How to Stay Healthy*, Greenwich, CT: Devin-Adair Publishers, 1990.

Palmer, B, *Body Weather*, Harrisburg, PA: Stackpole Books, 1976.

Pasichnyk, RM, *The Vital Vastness*, Volume 1, Lincoln, NE: Writer's Showcase, 2002.

Persinger, MA, *The Weather Matrix and Human Behavior*, New York: Praeger, 1980.

Petersen, WF, *Man, Weather, Sun*, Springfield, IL: Charles C Thomas, 1947.

Petersen, WF, *The Patient and the Weather*, Ann Arbor, MI: Edwards Brothers, 1935.

Reed, A, *Romantic Weather*, Hanover and London: University Press of New England, 1983.

Reiter, R, *Phenomena In Atmospheric And Environmental Electricity*, Amsterdam: Elsevier, 1992.

Rosen, S, *Weathering: How the Atmosphere Conditions your Body, your Mind, your Moods and your Health*, New York: M. Evans & Company Inc, 1979.

Rosenthal, N, *Seasons of the Mind: Why you Get the Winter Blues*, New York: Bantam Books, 1990.

Sargent, F, *Hippocratic Heritage: A History of Ideas About the Weather and Human Health*, New York: Pergamon Press, 1982.

Scheving, LE and Halberg, F, *Chronobiology: Principles and Applications to Shifts in Schedules*, Kluwer Academic, 1981.

Smith, A, *The Weather Factor – What is Happening to our Climate?*, London: Arrow, 2002.

Sobel, DS (ed), *Ways of Health - Holistic Approaches to Ancient and Contemporary Medicine*, New York: Harcourt Brace Jovanovich Inc, 1979.

Soyka, F and Edmonds, A, *The Ion Effect*, New York: Bantam Books, 1991.

Sulman, FG, *Health, Weather and Climate*, Basel, New York: S. KargerAG, 1976.

Sulman, FG, *The Effect of Air Ionisation, Electrical Fields, Atmospherics and Other Electric Phenomena on Man and Animal*, Springfield, IL: Charles C Thomas, 1980.

Thomson, WAR, *A Change of Air*, London: Adam & Charles Black, 1979.

Tromp, SW, *Biometeorology: The Impact of the Weather and Climate on Humans and their Environment*, London: Heyden & Son, 1980.

Tromp, SW, *Medical Biometeorology*, Amsterdam: Elsevier, 1963.

Tromp, SW, *Psychical Physics*, New York: Elsevier, 1949.

Watson, L, *Heaven's Breath: A Natural History of the Wind*, London: Sceptre, 1988.